W9-AZG-405

RESEARCH ON TECHNOLOGICAL INNOVATION, MANAGEMENT AND POLICY

Volume 4 • 1989

RESEARCH ON TECHNOLOGICAL INNOVATION, MANAGEMENT AND POLICY

A Research Annual

Editor: RICHARD S. ROSENBLOOM
Graduate School of Business Administration
Harvard University

Co-Editor: ROBERT A. BURGELMAN
Graduate School of Business
Stanford University

VOLUME 4 • 1989

JAI PRESS INC.

Greenwich, Connecticut *London, England*

55 Old Post Road, No. 2
Greenwich, Connecticut 06830

JAI PRESS LTD.
3 Henrietta Street
London WC2E 8LU
England

ISBN: 0-89232-798-7

Manufactured in the United States of America

CONTENTS

LIST OF CONTRIBUTORS

Paul S. Adler	Department of Industrial Engineering and Engineering Management Stanford University
Reinhard Angelmar	INSEAD France
Robert A. Burgelman	Graduate School of Business Stanford University
Yves Doz	INSEAD France
Raghu Garud	Carlson School of Management University of Minnesota
Michael Gibbons	Department of Science and Technology Policy University of Manchester England
J. Stanley Metcalfe	Department of Economics University of Manchester England
Gary Pisano	Graduate School of Business Administration Harvard University
C. K. Prahalad	School of Business Administration University of Michigan

Richard S. Rosenbloom — Graduate School of Business Administration, Harvard University

David J. Teece — School of Business Administration, University of California, Berkeley

Andrew H. Van de Ven — Carlson School of Management, University of Minnesota

INTRODUCTION:

EVOLUTIONARY PERSPECTIVES ON TECHNOLOGY, STRATEGY, AND ORGANIZATION

This book is the fourth in a series reporting on current thought and research on the management of technology. The papers in this volume report research at the intersection of technology, competitive strategy, and organization theory. Several papers share a concern with evolutionary processes shaping technology strategy, and focus either at the organizational or industry level of analysis.

The volume opens with a chapter by Burgelman and Rosenbloom providing an evolutionary process perspective on technology strategy. This chapter indicates how the evolution of a firm's technology strategy can be understood in terms of internal and external generative and integrative mechanisms. The paper discusses key dimensions of the substance of technology strategy and suggests ways to assess the comprehensiveness and integration of a technology strategy.

The second chapter, a bibliographical essay by Adler, provides a wide ranging overview, categorization, and synthesis of the different and somewhat dispersed literatures that bear on the management of technology. Adler discusses the threads that run through each of the main bodies of literature and begins to identify the linkages between them. A useful feature of this chapter is that it attempts to clarify the limits of and the lacunae in our current knowledge, and thus serves to highlight broad areas where further research would be fruitful.

The next two chapters are at the industry level of analysis. The chapter by Metcalfe and Gibbons outlines a variation-selection-retention framework for analysis of the dynamic relationships between technology and long-run competitive performance. This chapter raises the important question of how new technological forms are created, and how these technological forms come to acquire economic significance. Squarely rooted in an ecological perspective on organizations, Metcalfe and Gibbons argue that attention of scholars should be shifted from the search for ideal types toward the search for distributions of phenomena. An important conclusion from their formal analysis is that, under certain conditions, inferior technology may come to dominate the market. They view organizations as playing a key role in generating variety and carrying technological performance through time in a structured, positive way, and consider the manner in which organizations achieve their role as both constrained and stimulated by their environment.

Van de Ven and Garud provide a framework for understanding the emergence of new industries going beyond the traditional industrial economic view. These authors conceptualize an emerging industry as a social system and identify and discuss three loosely-coupled hierarchical subsystems: instrumental, resource procurement, and institutional. Their discussion is cast in terms of an accumulation theory of change that includes overlapping periods of initiation, startup, and takeoff. Understanding the emergence of industries requires examining the process not only at the system level, looking at the industry as a whole, but also at the level of individual firms within the industry.

The final two chapters focus on analysis at the level of the firm. Combining elements from evolutionary theory and transaction cost analysis, Pisano and Teece examine the role of interorganizational collaborative relationships in a firm's technology strategy. They suggest that different inter-firm governance structures will be appropriate for different types of collaborative activities. Examining both relationships involving equity participation and those strictly governed by contracts, they apply their framework to the telecommunications equipment industry. Their empirical results show that the governance of collaborative arrangements is indeed related to the function of collaboration. This may have important normative implications for technology strategy.

In the final chapter, Prahalad, Doz, and Angelmar raise the important question of how the scope of a new technology may evolve in the context of a multibusiness firm. Using several case studies, they show how new technology may affect the boundaries of and interfaces between existing business units. This leads them to examine such top management issues as how to define the core technology of the firm and influence the shifting interdependences between business units as the core technology evolves.

Robert A. Burgelman
Richard S. Rosenbloom
Editors for Volume 4

TECHNOLOGY STRATEGY:
AN EVOLUTIONARY PROCESS PERSPECTIVE

Robert A. Burgelman and Richard S. Rosenbloom

1. INTRODUCTION

This chapter concerns itself with the strategic management of technology. In our view, technology is a resource that is as pervasively important in the organization as are financial and human resources. Its management is a basic business function. Viewing technology as a functional capability implies the need to develop a technology strategy, analogous to financial and human resource strategies. For technology, such a strategy is defined by a set of interrelated decisions encompassing, among others, technology choice, level of technology competence, level of funding for technology development, timing of technology introduction in new products/services, and organization for technology application and develop-

Research on Technological Innovation, Management and Policy
Volume 4, pages 1–23

ISBN: 0-89232-798-7

ment (e.g., Maidique and Patch, 1978). Technology strategy conceived in this fashion is much broader than R&D strategy (Mitchell, 1985).

Technology, itself, is defined here as the ensemble of theoretical and practical knowledge, knowhow, skills and artifacts that are used by the firm to develop, produce and deliver its products and services. Technology can be embodied in people, materials, facilities, procedures, and in physical processes. Key elements of technology may be implicit, existing only in embedded form (e.g., trade secrets based on knowhow). "Craftsmanship" and "experience" usually have a large tacit component, so that important parts of technology may not be expressed or codified in manuals, routines and procedures, recipes, rules of thumb or other explicit articulations (Burgelman et al., 1988). In the broadest sense, a firm's "technology" encompasses the entire set of technologies employed in the sequence of activities that constitute its value chain (Porter, 1985). This set can be decomposed for analysis into constituent elements according to type of technology, role in the value chain, or contribution to end product (or service) categories.

In the so-called "hi-tech" firms, technology is usually a major force driving the firm's competitive strategy. But all firms have in recent years become more aware of the critical role of technology in strategic decisions, and of the need to integrate technology strategy into the strategic management process. As a consequence, technology increasingly is recognized as an important element of business definition and competitive strategy. Abell (1980) identifies technology as one of three principal dimensions of business definition. As he notes, "technology adds a dynamic character to the task of business definition, as one technology may more or less rapidly displace another one over time." Porter (1983) observes that technology is among the most prominent factors that determine the rules of competition. Friar and Horwitch (1985) explain the growing prominence of technology as the result of historical forces: disenchantment with strategic planning, success of hi-tech firms in emerging industries, the surge of Japanese competition, recognition of the competitive significance of manufacturing, and the emergence of an academic interest in technology management.

The present chapter elaborates the view of technology as a functional capability, and develops an evolutionary process framework for discussing technology strategy. The perspective is that of the general manager responsible for the overall business strategy rather than that of the functional manager. The analysis, however, may help the functional-level technology manager think about his or her approach to the job. The framework focuses on technology strategy, but emphasizes its links with the overall business strategy.

The remainder of this chapter is organized as follows. The next section discusses an evolutionary process perspective on strategy-making and technology strategy. The following section examines the forces shaping technology strategy and provides examples of some companies' technology strategies. This is followed by a discussion of the main substantive dimensions of technology strategy and of the ways in which it is enacted. A brief final section summarizes the main themes of the paper.

II. TECHNOLOGY STRATEGY: AN EVOLUTIONARY PROCESS PERSPECTIVE

Paul Adler's chapter in this volume provides an overview of the growing literature on technology strategy and suggests the need for developing conceptual frameworks that would allow for cumulative knowledge to be generated. An evolutionary process perspective may be useful for this purpose. This perspective is emerging in economics (Nelson and Winter, 1982), organization theory (e.g., Weick, 1979; Aldrich, 1979; Hannan and Freeman, 1984) and strategic management (e.g., Burgelman, 1984; 1988b). It focuses on variation, selection, retention mechanisms for explaining dynamic behavior over time.

Gould (1987) has warned against the fallacy of unwarranted analogy in applying concepts from biological evolution to processes of cultural evolution. Taking into account Gould's admonition, the framework of evolutionary theory has nevertheless been fruitfully applied in the cultural evolutionary perspective that has emerged in recent years (e.g., Campbell, 1979; Weick, 1979; Boyd and Richerson, 1985). Gould's (1987) own interpretation of the establishment of QWERTY (David, 1985) as the dominant, if inferior, approach to laying out keys on typewriter keyboards in terms of the "panda thumb principle," shows the power of evolutionary reasoning in identifying and elucidating interesting phenomena concerning technological evolution.

Cultural evolutionary theory recognizes the importance of history, irreversibilities, invariance and inertia in explaining the behavior of social systems. But it also considers the effects of individual and social learning processes. A cultural evolutionary perspective may be useful for integrating extant literatures on technology. The study of technological development, for instance, contains many elements that seem compatible with the variation-selection-retention structure of evolutionary theory (e.g., Rosenberg, 1979; Krantzberg and Kelly, 1978; Abernathy, 1978; Clark, 1985). In fact, several other chapters in this volume adopt an evolutionary perspective. Metcalfe and Gibbons present an application of the evolutionary perspective at the level of populations of technologies. While Van de Ven and Garud criticize applications of a Darwinian view of evolution to the emergence of new industries around new technologies, their analysis emphasizes the social context of innovation and appears consistent with a cultural evolutionary perspective.

A. Capabilities-Based Perspective

Embedded in the cultural evolutionary perspective is a view of strategy-making as a social learning process. In this view, strategy is inherently a function of the quantity and quality of organizational capabilities. Organizational capabilities are the source of opportunities which are discovered, selected, and retained in the strategy-making processes. Experience with (''performing'') a strategy is expected to have feedback effects on the set of organizational capabilities. The general structure of the capabilities-based perspective (Burgelman, 1984; 1988a) is presented in Figure 1.

It is useful to note several key ideas underlying this framework. One is that successful firms develop distinctive strategies in the course of their development and that the direction of this development cannot be completely determined at the outset. A second key idea is that increasing the firm's capabilities is a mechanism for stimulating strategic development. This is consistent with Itami's (1987) discussion of the accumulation of ''invisible assets'' as a key factor in an organization's development. It is also consistent with Maidique and Zirger's (1985) study of the product learning cycle. A third idea is that ''unlearning'' is an important aspect of organizational learning (Imai, Nonaka and Takeuchi, 1985; Levitt and March, 1988).

Within this perspective, ''performance'' is viewed in terms of experience with actually performing the different tasks involved in carrying out a strategy. This view of performance is akin to the use of the term in craftsmanship, the arts, and sports. Studying the details of actually performing a strategy may shed light on exactly how skills are accumulated and how organizational learning and unlearning, in their various forms actually come about.

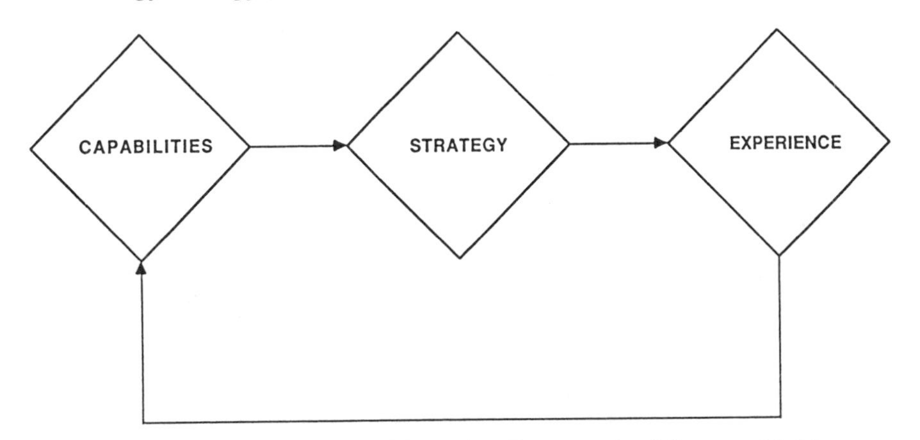

Figure 1. An Organizational Learning Framework of Strategy-Making

III. FORCES SHAPING TECHNOLOGY STRATEGY

The factors shaping technology strategy comprise a number of varied forces, mediating the influence of capabilities and experience. In this section, we explore a simple framework—depicted in Figure 2—which augments the theoretical context for the multilevel analysis of technology proposed by Rosenbloom (1978). This concept differentiates the factors according to locus—within or outside the firm—and functions—generative or integrative—in the strategy process. We hope that this classification might help to focus research and enable managers to examine their firms' strategies within a theoretical context.

In summary, the idea expressed in Figure 2 is that technology strategy emerges from organizational capabilities, shaped by the generative forces of the firm's strategic behavior and evolution of the technological environment, and by the integrative mechanisms of the firm's organizational context and the environment of the industry in which it operates. Each of these mechanisms is briefly addressed in what follows.

1. Internal Forces

Generative Mechanisms: Strategic Behavior. Strategy-making with respect to technology is a subset of the broader strategy-making of the firm. From an evolutionary perspective, a firm's concept of strategy

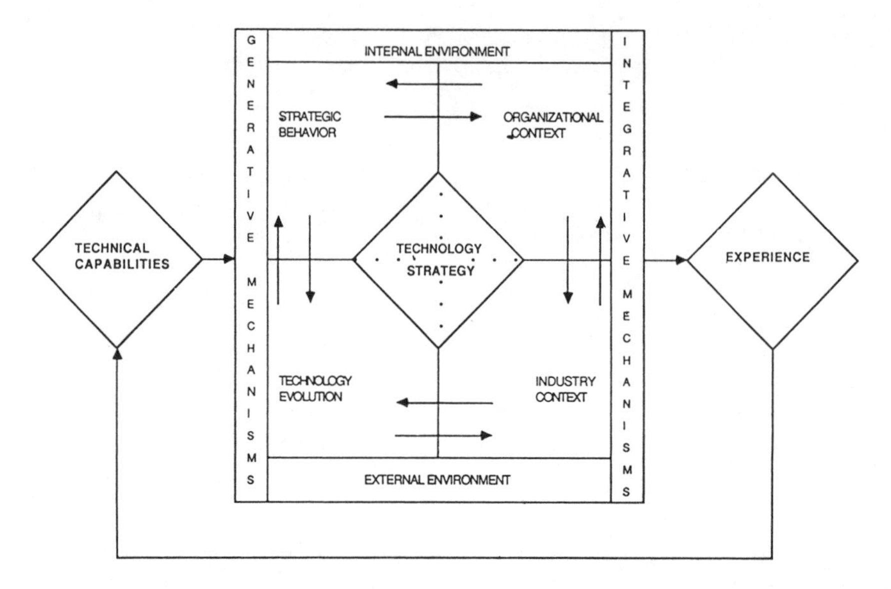

Figure 2. An Evolutionary Process Framework for Technology Strategy

impounds the social learning about the distinctive competences on which past success is based (Selznick, 1957; Burgelman, 1988b). McKelvey and Aldrich (1983) view distinctive competence, "comps," as "the combined workplace (technological) and organizational knowledge and skills . . . that together are most salient in determining the ability of an organization to survive" (p. 112). Nelson and Winter (1982), in a similar vein, use the concept of "routines," which they consider to play a role similar to that of genes in biological evolution.

The analogy between genes and distinctive competence (capabilities) raises the issue of where strategic change and new technical capabilities, ultimately, come from. It seems plausible to expect that the "comps" and "routines" that have evolved in the course of achieving organizational survival will be naturally applied over and over again in the course of the firm's strategic action. To a large extent strategic behavior is, indeed, *induced* by the prevailing concept of strategy of the firm and results, through new product and process development efforts falling within its scope, in the enhancement and augmentation of *existing* technical capabilities. The induced strategic process is likely to manifest a degree of inertia relative to the cumulative changes in the external environment (Hannan and Freeman, 1984; Burgelman, 1988b). Cooper and

Schendel (1978), for instance, found that established firms, when confronted with the threat of radically new technologies, were likely to increase their efforts to improve the existing technology rather than switch to the new technology even after the latter had passed the threshold of viability.

But firms will usually also exhibit some amount of autonomous strategic behavior, aimed at getting the firm into new areas of business (e.g., Penrose, 1968; Burgelman, 1983). These initiatives often are rooted in technology development efforts. In the course of their work, for example, technical people may serendipitously discover results that provide the basis for redirection or replacement of major technologies of the firm. The existence of a corporate R&D capability often provides a substratum for the emergence of such new technical possibilities (Rosenbloom and Kantrow, 1982; Burgelman and Sayles, 1986).

Participants engaging in autonomous strategic behavior serve to explore the boundaries of a firm's capabilities and corresponding opportunities sets (Burgelman, 1983). As Itami (1983) has observed, "in reality, many firms do not have . . . complete knowledge and discover the full potential of their ability only after the fact" (p. 15). The degree to which autonomous strategic behavior can be exploited, however, depends critically on the organizational context.

Integrative Mechanisms: Organizational Context. Internal context encompasses administrative (Bower, 1970) and cultural (Ouchi, 1980) factors which affect participants' expectations about which types of strategic behavior are likely to be supported in the organization. From an evolutionary point of view, organizational context serves as an internal selection mechanism (Burgelman, 1988) affecting the strategic management capacity of the firm, that is, (1) the firm's ability to exploit opportunities associated with its current strategy (induced process), (2) its ability to take advantage of opportunities that emerge spontaneously outside the scope of the current strategy (autonomous process), and (3) its ability to balance the different modes of strategic action associated with (1) and (2) over time (Burgelman and Sayles, 1986).

The evolutionary process perspective draws attention to the fact that organizational context takes shape over time, and reflects the dominant culture (values) of the firm. The dominant culture as it relates to technology may be different depending on whether the firm's distinctive competences are rooted in science (e.g., pharmaceutical firms), engineering (e.g., semiconductor firms) or manufacturing (e.g., Japanese firms);

whether the product development process has been driven by technology push, need pull or a more balanced approach; whether operations are viewed as strategic or not; and so on.

2. *External Forces*

Generative Mechanisms: Technology Evolution. A firm's technology strategy is rooted in the evolution of its technical capabilities. However, the dynamics of these capabilities, and hence the technology strategy, are not completely endogenous. A firm's technical capabilities are affected in significant ways by the evolution of the broader areas of technology of which they are part. Different aspects of technological evolution have been discussed in several studies: (1) technology regimes and their development along a particular trajectory (e.g., Twiss, 1982; Foster, 1986); (2) the emergence of dominant designs, design hierarchies, industry standards given a technological regime (Abernathy, 1978; Clark, 1985; and Metcalfe and Gibbons in this volume); (3) the interplay between product and process technology development within design configurations under a particular technological regime (Abernathy, 1978); (4) the emergence of new technological regimes and their trajectories (S-curves) (Foster, 1986); (5) the capability (competence) enhancing or destroying nature of new technological regimes (Astley, 1985; Tushman and Anderson, 1986); (6) dematurity (Abernathy, Clark and Kantrow, 1983); and (7) the locus of origin of new technological regimes (von Hippel, 1988). These studies highlight some of the major evolutionary forces associated with technological development which, as noted earlier, affect and are affected by the strategic behavior within firms.

Integrative Mechanisms: Industry Context. One important aspect of industry context is the competitive market place. Competitive strategy and the quest for competitive advantage take into account five major forces (Porter, 1980): (1) rivalry among existing firms, (2) bargaining power of buyers, (3) bargaining power of suppliers, (4) threat of new entrants, and (5) threat of substitute products or services. The interplay of these five forces is expected to determine the appropriate content of competitive strategy. Technological change affects each of the five forces (Porter, 1983) and technology strategy may serve as a potentially powerful tool for pursuing the generic competitive strategies (overall cost leadership, overall differentiation, focused cost leadership, focused differentiation).

Contextual factors affect the choice of leadership versus followership (Porter, 1983). Examining this choice further, Teece (1986) argues that the fundamental building blocks of a decision framework include: (1) the appropriability regime associated with a technological innovation, (2) complementary assets needed to commercialize a new technology, and (3) the dominant design paradigm. The interplay of these factors affects the likely distribution of profits generated by a technological innovation among the different parties involved as well as the strategic choices concerning the optimal boundaries of the innovating firm's capabilities set.

The chapter by Metcalfe and Gibbons in this volume shows how the economic aspects of the industry context can be operationalized from an evolutionary perspective. Industry context selects among technological alternatives and the most efficient technological alternative need not necessarily become dominant. Still another important aspect of industry context is the social embeddedness of economic transactions. The chapter by Van de Ven and Garud in this volume examines the broader institutional forces associated with the industry context.

3. *Applying the Framework: Some Examples*

Before moving on to discuss the implications of the evolutionary perspective for the substance and deployment of technology strategy, it seems appropriate to examine how the framework may help gain insight in a firm's technology strategy. Several examples can serve this purpose.

- Crown Cork and Seal has been able to do very well over a 30-year period as a relatively small player in a mature industry. Technology strategy seems to have contributed significantly to Crown's success. When taking over the company in 1957, CEO John Connelly recognized the existence of Crown's strong skills in metal formation (capabilities) and built on these to specialize in "hard to hold" applications for tin cans. He developed strong relations with the steel companies and convinced them to quickly adapt a major external technological innovation (*technology evolution*)—the two-piece can—initiated by an aluminum company for use with steel. He did not want Crown to have an R&D department, but developed strong links between a highly competent technical sales force and an applications-oriented engineering group to be able to provide complete technical solutions for customers' "filling needs"—a dominant value at CC&S (*organizational context*). Over the years,

CC&S continued to stick to what it could do best (*strategic behavior*) while its competitors were directing their attention to diversification and gradually lost interest in the metal can industry (*industry context*). CC&S has continued to refine the skill set that made it the only remaining independent metal can company of the four original major players (*experience*).

• Marks and Spencer (M & S), a British retailer with a worldwide reputation for quality, and Bank One Corporation, a Midwest banking group that consistently ranks among the most profitable U.S. banking operations, are two companies in the services sector that have used technology effectively to gain competitive advantage.

The success of M&S is based on a consistent strategy founded on an unswerving commitment to giving the customer "good value for money." The genesis of its technology strategy was the transformation, in 1936, of a small textile testing department into a "Merchandise Development Department," designed to work closely with vendors to bring about improvements in quality. Thirty years later, the M & S technical staffs, then numbering more than two hundred persons working on food technology as well as textiles and home goods, allowed M&S, quite literally, to control the cost structure of its suppliers. The development of the technical capability itself was driven by the strong value of excellent supplier relationships held and continuously reinforced by top management.

In 1958, the new CEO of City National Bank of Columbus Ohio (CNB), John G. McCoy, persuaded his Board to invest 3% of profits each year to support a "research and development" activity. Over the next two decades, CNB, which became the lead bank of Banc One Corporation, developed capabilities that made it a national leader in the application of electronic information-processing technologies to retail banking. It was the first bank to install automatic teller machines and a pioneer in the development of bank credit cards, point-of-sale transaction processing, and home banking. While not all of its innovative ventures succeeded, each contributed to the cumulative development of a deep and powerful technical capability that remains a distinctive element of the bank's highly successful competitive strategy.

The three companies cited above are notable for the consistency of their strategic behavior over several decades. The following example illustrates the problems that can arise in a time of changing technology and industry context when a fundamental change in strategy is not matched by corresponding adaptation of the organizational context.

• The National Cash Register Company (NCR) built a dominant position in world-wide markets for cash registers and accounting machines on the basis of superior technology and an outstanding salesforce created by the legendary John H. Patterson. By 1911, NCR had a 95% share in cash register sales. Scale economies in manufacturing, sales, and service presented formidable barriers to entry to its markets (*industry context*), preserving its dominance for another 60 years. Highly developed skills in the design and fabrication of complex low-cost machines (*capabilities*) not only supported the strategy, they also shaped the culture of management, centered on Dayton where a vast complex housed engineering and fabrication for the traditional product line (*organizational context*). In the 1950s, management began to build new capabilities in electronics (*a revolution, not just evolution in technology*) and entered the emerging market for Electronic Data Processing (EDP). A new strategic concept tried to position traditional products (registers and accounting machines) as "data-entry" terminals in EDP systems (*changing strategic behavior*). But a salesforce designed to sell stand-alone products of moderate unit cost proved ineffective in selling high-priced "total systems." At the same time, the microelectronics revolution destroyed the barriers inherent in NCR's scale and experience in fabricating mechanical equipment. A swarm of new entrants found receptive customers for their new electronic registers (*changing industry context*). As market share tumbled and red-ink washed over the P&L in 1972, the chief executive was forced out. His successor was an experienced senior NCR manager who had built his career entirely outside of Dayton. He moved swiftly to transform the ranks of top management, decentralize manufacturing (reducing employment in Dayton by 85%), and restructure the salesforce along new lines. The medicine was bitter, but it worked; within 2 years, NCR had regained leadership in its main markets, and was more profitable in the late 1970s than it had been at any point in the 1960s.

IV. TECHNOLOGY STRATEGY: SUBSTANCE AND ENACTMENT

A strategic view of technology has many facets, but the substance of a firm's technological strategy can be grasped in terms of a few fundamental themes. In essence, a business uses technology to create enhanced value in its products or services and to gain sustainable advantage in relation to its rivals. Two important dimensions of strategy, then, are the way it positions the business competitively and in relation to the value chain. But the implications of a particular posture in these respects can be various, depending upon choices in two other dimensions, those defining the scope and depth of technological resources. In this section we first consider, briefly, each of these four dimensions of the substance of technology strategy. We then examine the notion of performance, the ways that strategy is realized in practice. The section concludes with conjectures about the significance of comprehensiveness and integration.

1. *Substance of Technology Strategy*

Competitive Positioning. Technology strategy, viewed in competitive terms, is an instrument of a more comprehensive corporate strategy, as we have noted earlier. As part of the broader strategy, a business adopts a particular *competitive stance* toward technology, defining the role that technology should play in establishing generic competitive advantage: How *process* technology, *product* technology and technical *support* capabilities are used to achieve cost leadership or differentiation (Porter, 1985).

From an evolutionary perspective, however, a firm's competitive advantage is more likely to arise from the *unique* aspects of its strategy than from characteristics it shares with others. Companies that have been successful over long periods of time, such as Crown Cork and Seal, and Marks and Spencer, have developed capabilities that are quite distinct from those of their competitors and not easily replicable. The strategies of such companies cannot easily be classified simply in terms of differentiation *or* cost leadership; they combine both. The technical skills and capabilities that they have assembled and built into their organizations over the span of many years, and which they relentlessly hone and augment, are not available in the market and would take a long time to be replicated. The ability to maintain uniqueness that is salient in the market place, however, implies continuous alertness to what competitors are doing and should not be confused with insulation and an inward-looking orientation.

Technology, from a competitive point of view, can be used in a defensive role, sustaining achieved advantage in product differentiation or cost, or, offensively, as an instrument of expansion, to create new advantage in established lines of business or to open the door to new products and markets. The implications of "technological leadership" have been explored in earlier writings on technology and strategy (e.g., Ansoff and

Stewart, 1967; Maidique and Patch, 1979). As noted earlier, Porter (1981) relates this concept more directly to a broader framework of competitive advantage, identifying conditions under which leadership is likely to be rewarded in the market place. Teece (1986) extends the analysis by identifying the importance of appropriability regimes and control of specialized assets. These discussions define leadership in terms of the timing (relative to rivals) of commercial use of new technology. A broader strategic definition would identify "leadership" in terms of relative advantage in the command of a body of technology. This sort of leadership results from commitment to a "pioneering" role in the development of a technology (Rosenbloom and Cusumano, 1987), as opposed to a more passive "monitoring" role. Technological leaders thus have the capability to be first-movers, but may elect not to do so.

Looking at technological leadership from an evolutionary perspective draws attention to the importance of *accumulation* of capabilities as a competitive tool (Itami, 1987). Given the importance of history, irreversibilities, and inertia, leadership is not something that can easily be bought in the market or quickly "plugged" into the organization. Rather, it involves painstaking, patient and persistent building of technical capabilities in all areas that have been determined as having strategic importance for the future development of the firm, even though it may seem cheaper ("efficient") to rely on outsiders for supplying the capability. Thinking strategically about technology means raising the question of how a particular technical capability may affect a firm's future degrees of freedom and the control over its fate. Concretely, this involves identifying and tracking key technical parameters, considering the impact on speed and flexibility of product and process development as technologies move through their life cycles. This, in turn, requires distinguishing carefully between technologies that are common to all players in the industry and have little impact on competitive advantage and those that are proprietary and likely to have a major impact on competitive advantage. It also depends on paying attention continuously to new technologies that are beginning to manifest their potential for competitive advantage and those that are as yet only beginning to emerge (A.D. Little, 1981; Booz-Allen and Hamilton, 1981; Burgelman et al., 1988).

Technology and the Value Chain. A second dimension of the substance of technology strategy defines a value chain stance toward the use of technology: that is, how technology is to be used to add value in process, products or technical service delivered to customers or users. A firm that lacks important capabilities in a given constituent technology (by choice, or otherwise) may elect to embrace it through supply relationships or strategic collaborations (see the chapter by Pisano and Teece in this volume). Hence the value chain stance also answers the question of how value-creating technology is sourced—in-house or outside the boundaries of the firm.

From an evolutionary process perspective, a key issue is how technology could affect the rate of *improvement* in performing the tasks involved in each of the stages of the value chain. Continuous concern with improvement in all aspects of the value creation and delivery process will guard the firm against quirky moves that endanger the accumulated learning in the organization. Similarly, viewing the issue of sourcing from this perspective highlights the importance of managing interdependencies with external providers of capabilities. One requirement of this is a continuous concern for gaining as much *learning* as possible from the relationship in terms of capabilities and skills rather than a concern solely with price. To the extent that a firm engages in alliances, it seems necessary to establish the requisite capabilities for managing the relationships. As noted earlier, Marks and Spencer's strong technical staff made it possible to have unique and valuable supplier relationships.

Scope of Technology Strategy. A third dimension of the substance of technology strategy defines the scope of technologies actively attended to by the firm. In one respect this goes back to the notion of the value chain and the concern with technologies in its different stages. Firms may concern themselves to a different degree with all of these. No firm, however, can hope to operate on the frontiers of all technologies relevant to its operations. Thus, it seems more useful to limit the scope of technology strategy to the set of technologies considered by the firm to have a material impact on its competitive advantage. This set of technologies can be called the *core technology;* other technologies, then, are *peripheral.* Of course, in a dynamic world, peripheral technologies today may become core technologies tomorrow and vice versa. This dimension of technology strategy is especially important in relation to the threat of new entrants in the firm's industry. All else equal, firms covering more areas of technology in their core would seem to be less *vulnerable* to attacks from new entrants attempting to gain position through producing and delivering new types of technology-based customer value. Of course, resource constraints will put a limit on how many technologies the firm can opt to develop internally.

Defining the scope of the technology strategy determines which ones within the entire range of technologies employed by the firm are placed within the core. These are the areas in which the firm needs to assess its distinctive technological competences and to decide whether to be a technological leader or follower, first-mover or second-to-market, and whether to develop requisite capabilities in-house or through vendors and allies. Such assessments and decisions identify an array of targets for technology development. The irreversibility of investments in technology makes the choice of these targets an especially salient dimension of strategy. For any given technical field, moreover, both risk and return will vary between product and process development and also according to the degree of novelty inherent in the development. Targeted technology development may range from minor improvements in a mature process to the employment of an emerging technology in the first new product in a new market (Rosenbloom, 1985).

The scope of a firm's technology strategy may be determined to a significant extent by its scale and business focus. Businesses built around large, complex systems like aircraft, automobiles, or telecommunication switches demand the ability to apply and integrate numerous distinct types of expertise creating *economies* of scope and/or scale, and *synergies*. General Electric, for instance, was reportedly able to bring to bear high powered mathematical analysis, used in several divisions concerned with military research on submarine warfare, to the development of computerized tomography (CT) products in its medical equipment division (Rutenberg, 1986 p. 28–9.). Other fields may actually contain diseconomies of scale, giving rise to the popularity of "skunkworks" and discussion of the "mythical man-month" (Brooks, 1975). The chapter by Prahalad, Doz, and Angelmar in this volume examines how the emergence of a new technology raises issues concerning the scope of the innovation that the firm should pursue and how this, in turn, may impact the delineation of the set of core technologies of the firm and the boundaries between the business units that it comprises.

Depth of Technology Strategy. The fourth dimension of the substance of technology strategy concerns the depth of its prowess within the core technology. This prowess can be expressed in terms of the number of technological options that the firm has available. The depth of a firm's technological strategy is determined to a significant extent by the intensity of its resource expenditures. The variation among manufacturing firms is pronounced: many firms do not spend on R&D; a few commit as much as 10% of revenues to it. While inter-industry differences can explain a large share of this variation, substantial differences in R&D intensity still remain between rivals.

From an evolutionary perspective, depth of technology is likely to be correlated with the firm's capacity to anticipate technological developments in particular areas early on. It provides the basis for acting in a *timely* way. As noted earlier, however, being able to take advantage of this lead in knowledge in terms of developing new products or services will to a large extent depend on the organizational context. Firms that can organize themselves for attaining greater technological depth may also benefit from increased *flexibility* to respond to new demands from customers/users. Imai et al. (1985) for instance, describe how Japanese firms' use of multiple layers of contractors and subcontractors in an external network allows extreme forms of specialization in particular skills which, in turn, provides both flexibility and speed in response and the potential for cost savings, since the highly specialized subcontractors operate on an experience curve even at the level of prototypes.

2. *Performance*

The enactment of a strategy is embodied in the performance of the tasks that make it real. For technology, the main processes concern its acquisition, development, and support.

Acquisition. Since the sources of technology are inherently varied, so must be the mechanisms employed to make it accessible within the firm. But it is useful to make a simple distinction between inventive activity within the firm and acquisitive functions which import technology originating outside the firm.

Relatively few firms—primarily the largest ones in the R & D intensive industries—find it useful to support the kind of science-based R & D that can lead to important new technologies. But those firms, in aggregate, shape the technological directions of the economy, so their strategies are especially salient. Each firm's strategy finds partial expression in the way it funds, structures, and directs the R & D activities whose mission is to create new pathways for technology. A current example of the importance of this function is the emergence of high-temperature superconductivity, based on a discovery made in an IBM research laboratory.

Every firm finds that it must structure ways to acquire certain technologies from others. The choices made in carrying out those tasks can tell us a great deal about the underlying technology strategy. To what extent does the firm rely on on-going alliances, as opposed to discrete transactions for the acquisition of technology (Hamilton, 1986)? Is the acquisition structured to create the capability for future advances to be made internally, or will it merely reinforce dependence? For example, when Japanese electronics firms acquired technology from abroad during the 1950s and 1960s most of them structured it in ways that made it possible for them to become the leaders in pushing the frontiers of those same technologies in the 1970s and 80s.

Development. Strategy is also acted out in the functions which develop technology as products and processes. The character of product development activities embodies important aspects of the scope and depth of technology strategy and its competitive role. An understanding of what the strategy is can be gained from considering the level of resources committed, the way they are deployed, and how they are directed. How does the organization strike the delicate balance between letting technology drive development and using the marketplace to drive technology? The mammoth personal computer industry was founded when the young engineers of Apple sought to exploit the potential inherent in the microprocessor, at a time when corporate managers in Hewlett-Packard, IBM, and others were disdainful of the commercial opportunity. A decade later, however, it is clear that market needs are now the primary force shaping efforts to advance the constituent technologies. In a number of industries—autos and semiconductors would be two examples—competitiveness rests substantially on design and manufacturing process capabilities. There strategy is embodied in the links between product and process engineering, in the resources committed to capital equipment, in the concern for quality, and similar aspects of this domain.

Support. The function commonly termed "field service" creates the interface between the firm's technical function and the users of its products or services. Experience in use provides important feedback to enhance the capabilities which generate the technology (Rosenberg, 1982, Chapter 8). Airline operations, for example, are an essential source of information about jet engine technology. The technology strategy of a firm, then, finds important expression in the way it carries out this important link to users. Two-ways flows of information are relevant: expert knowledge from product developers can enhance the effectiveness of field operations, while feedback from the field informs future development.

MODES OF EXPERIENCE / SUBSTANCE	EXTERNAL TECHNOLOGY SOURCING	INVENTIVE ACTIVITY	PRODUCT DESIGN DEVELOPMENT	PROCESS DESIGN DEVELOPMENT	TECHNICAL SUPPORT SERVICE
RIVALRY STANCE					
VALUE CHAIN STANCE					
SCOPE					
DEPTH					

Figure 3. Technology Strategy: Comprehensiveness and Integration

3. *Two Conjectures*

Implicit in the foregoing discussion are two normative conjectures about technology strategy. The first is that the substance of technology should be comprehensive. That is, strategy, as it is enacted through the various tasks of acquisition, development, and support, should address all four substantive dimensions and do so in ways that are consistent across the dimensions. Second, we suggest that strategy should be integrated with operations. That is, that each of the tasks should be informed by the positions taken on the four substantive dimensions, in ways that create consistency across the various tasks. The matrix illustrated in Figure 3 is intended to suggest a framework for mapping the interactions among these factors and assessing the degree of comprehensiveness and integration in a specific situation.

V. CONCLUSION

This chapter argues that an evolutionary perspective provides a useful framework for thinking about the nature of technology strategy and about its role in the broader competitive strategy of a firm. The essence of this perspective is that strategy is built on capabilities and tempered by experience. These three main constructs—capabilities, strategy, and experience—are tightly interwoven in reality. Capabilities give strategy its force; strategy enacted creates experience that modifies capabilities.

Thus, we argue, it is useful to examine the technological strategy of a firm through the lenses of the underlying capabilities and the resultant modes of performance. Central to this idea is the notion that the reality of a strategy lies in its enactment, not in those pronouncements that appear to assert it. Through these lenses, we suggest, one can readily discern the scope and depth of the strategy and the way that it positions the firm in relation to rivals in the marketplace and in relation to others in the value chain. In other words, the substance of strategy can be found in its performance in the various modes by which technology is acquired and deployed—sourcing, development, and support. The ways in which these tasks are actually performed, and the ways in which their performance contributes, cumulatively, to capabilities convey the real substance of strategy.

A second central idea in this paper is that the on-going interactions of capabilities-strategy-experience occur within a matrix of generative and integrative mechanisms that shape strategy. These mechanisms (sketched in Figure 2) are both internal and external to the firm. Anecdotal evidence suggests that successful firms operate within some sort of harmonious equilibrium of these forces. Major change in one, as in the emergence of a technological discontinuity, ordinarily must be matched by adaptation in the others.

Which leads to the final conjecture of this paper, namely, that it is advantageous to attain a state in which technology strategy is both comprehensive and integrated. By comprehensive we mean that it embodies consistent answers to the issues posed by all four substantive dimensions. By integrated we mean that each of the various modes of performance is informed by the strategy.

These ideas suggest several areas where further research could be fruitfully undertaken. Longitudinal field research at the level of individual organizations could shed light on the behavioral demands of how an organization develops distinctive technological capabilities, and how these are enhanced, augmented, refocused or dissipated through internal ventures, acquisitions, alignments and spin-offs. Of particular interest are questions about how the organization builds a business strategy on its evolving capabilities and skills and how, in turn, the business strategy and its specific enactment may facilitate or impede the further development of technological capabilities and skills. The relationships between organizational growth and capabilities development would also appear to be an interesting target for further research. Comparative longitudinal research at the level of industries may shed light on why technology strategies vary among the firms they contain and may document further the forces affecting different firms' modes of adaptation.

ACKNOWLEDGMENTS

The authors gratefully acknowledge support provided by the Strategic Management Program, Graduate School of Business, Stanford University, and the Division of Research of the Harvard Business School.

REFERENCES

Abell, D., 1980. *Defining the Business.* Englewood Cliffs, NJ: Prentice-Hall.

Abernathy, W. J., 1978. *The Productivity Dilemma: Roadblock to Innovation in the Automobile Industry.* Baltimore: Johns Hopkins University Press.

Abernathy, W. J., and A. M. Kantrow, 1983. *Industrial Renaissance*. New York: Basic Books.

Aldrich, H. E., 1979. *Organizations and Environments*. Englewood Cliffs, NJ: Prentice-Hall.

Ansoff, H. I., and J. M. Stewart, 1967. "Strategies for a Technology-Based Business," *Harvard Business Review*, November–December.

Astley, W. G., 1985. "The Two Ecologies: Population and Community Perspectives on Organizational Evolution," *Administrative Science Quarterly*, 30, 224–241.

Booz-Allen and Hamilton, 1981. "The Strategic Management of Technology," *Booz-Allen & Hamilton: Outlook*, Fall–Winter, 1981.

Bower, J. L., 1970. *Managing the Resource Allocation Process*. Boston, MA: Division of Research, Graduate School of Business Administration, Harvard University.

Boyd, R., and R. Richerson, 1985. *Culture and the Evolutionary Process*. Chicago: University of Chicago Press.

Brooks, F. P., 1975. *The Mythical Man-Month*. Reading, MA: Addison-Wesley.

Burgelman, R. A., 1983. "Corporate Entrepreneurship and Strategic Management: Insights from a Process Study," *Management Science*, 29, 1349–1364.

Burgelman, R. A., 1986. "Strategy-Making and Evolutionary Theory: Towards a Capabilities-Based Perspective," in Tsuchiya, M. (ed.), *Technological Innovation and Business Strategy*, Tokyo: Nihon Keizai Shinbunsha.

Burgelman, R. A., 1988a. "Strategy-Making as a Social Learning Process: The Case of Internal Corporate Venturing," *Interfaces*, 18, 3, (May–June) 74–85.

Burgelman, R. A., 1988b. "Intraorganizational Ecology of Strategy-Making and Organizational Adaptation," Research Paper, Graduate School of Business, Stanford, CA: Stanford University, August.

Burgelman, R. A., and L. R. Sayles, 1986. *Inside Corporate Innovation*. New York: Free Press.

Burgelman, R. A., T. J. Kosnik, and M. Van den Poel, 1988. "Toward an Innovative Capabilities Audit Framework," in R. A. Burgelman and M. A. Maidique (eds.), *Strategic Management of Technology and Innovation*. Homewood, IL: Irwin.

Campbell, D. T., 1969. "Variation and Selective Retention in Sociocultural Evolution," *General Systems*, 14, 69–85.

Clark, K. B., 1985. The Interaction of Design Hierarchies and Market Concepts in Technological Evolution," *Research Policy*, 14, 235–251.

Cooper, A. C., and D. Schendel, 1976. "Strategic Responses to Technological Threats," *Business Horizons*, (February) 61–63.

David, P. A., 1985. "Clio and the Economics of QWERTY," American *Economic Review*, 75, 2, (May) 332–337.

Foster, R. N., 1986. *Innovation: The Attacker's Advantage*. New York: Summit.

Friar, J., and M. Horwitch, 1985. "The Emergence of Technology Strategy: A New Dimension of Strategic Management," in *Technology in Society*, 7, 2/3, 143–178.

Fusfeld, A., 1978. "How to Put Technology into Corporate Planning," *Technology Review*, (May).

Gould, S. J., 1987. "The Panda's Thumb of Technology," *Natural History*, (January) 14–23.

Hamilton, W. F., 1985. "Corporate Strategies for Managing Emerging Technologies," in *Technology in Society*, 7, 2/3, 197–212.

Hannan, H. T., and J. H. Freeman, 1984. "Structural Inertia and Organizational Change," *American Sociological Review,* 43, 149–164.

Imai, K., I. Nonaka, and H. Takeuchi, 1985. "Managing the New Product Development Process: How Japanese Learn and Unlearn," in K. B. Clark, R. H. Hayes, and C. Lorenz, *The Uneasy Alliance: Managing the Productivity-Technology Dilemma.* Boston, MA: Harvard Business School Press.

Itami, H., 1983. "The Case for Unbalanced Growth of the Firm," Research Paper Sei es #681, Graduate School of Business, Stanford, CA: Stanford University.

Itami, H., 1987. *Mobilizing Invisible Assets.* Cambridge, MA: Harvard University Press.

Kelly, P., and M. Kranzberg, eds., 1978. *Technological Innovation: A Critical Review of Current Knowledge.* San Francisco: San Francisco Press.

Levitt, B., and J. G. March, 1988. "Organizational Learning," *Annual Review of Sociology,* 14.

Little, A. D., 1981. "The Strategic Management of Technology," *European Management Forum.*

Maidique, M. A., and P. Patch, 1978. "Corporate Strategy and Technological Policy," *Harvard Business School,* Case #9-679-033, rev. 3/80.

Maidique, M. A., and B. J. Zirger, 1985. "New Products Learning Cycle," *Research Policy,* (December) 1–40.

McKelvey, B., and H. E. Aldrich, 1983. "Populations, Organizations, and Applied Organizational Science," *Administrative Science Quarterly,* 28, 101–128.

Mitchell, G. R., 1986. "New Approaches for the Strategic Management of Technology," in M. Horwitch (ed.), *Technology in Society,* 7, 2/3, 132–144.

Nelson, R. R., and S. G. Winter, 1982. *An Evolutionary Theory of Economic Change.* Cambridge, MA: Harvard University Press.

Ouchi, W., 1980. "Markets, Bureaucracies and Clans," *Administrative Science Quarterly,* 25, 129–141.

Penrose, E. T., 1980. *The Theory of the Growth of the Firm.* White Plains, NY: M. E. Sharpe.

Porter, M. E., 1980. *Competitive Strategy.* New York: Free Press.

Porter, M. E., 1983. "The Technological Dimension of Competitive Strategy," in R. S. Rosenbloom (ed.), *Research on Technological Innovation, Management and Policy,* 1, 1–33.

Porter, M. E., 1985. *Competitive Advantage.* New York: Free Press.

Prestowitz, C. V., Jr., 1988. *Trading Places.* New York: Basic Books.

Rosenberg, N., 1982. *Inside the Black Box.* Cambridge: Cambridge University Press.

Rosenbloom, R. S., 1978. "Technological Innovation in Firms and Industries: An Assessment of the State of the Art," in P. Kelly and M. Kranzberg (eds.), *Technological Innovation: A Critical Review of Current Knowledge.* San Francisco: San Francisco Press.

Rosenbloom, R. S., 1985. "Managing Technology for the Longer Term: A Managerial Perspective," in K. B. Clark, R. H. Hayes, and C. Lorenz, (eds.), *The Uneasy Alliance: Managing the Productivity-Technology Dilemma.* Boston, MA: Harvard Business School Press.

Rosenbloom, R. S., and A. M. Kantrow, 1982. "The Nurturing of Corporate Research," *Harvard Business Review,* (January–February), 115–123.

Rosenbloom, R. S., and M. A. Cusumano, 1987. "Technological Pioneering: The Birth of the VCR Industry,'' *California Management Review,* XXIX, 4, (Summer), 51–76.

Rutenberg, David, 1986. ''*Umbrella Pricing,*'' Working Paper, Queens University.

Selznick, P., 1957. *Leadership in Administration.* New York: Harper and Row.

Teece, D. J., 1986. ''Profiting from Technological Innovation: Implications for Integration, Collaboration, Licensing and Public Policy,'' *Research Policy,* 15, 285–305.

Tushman, M. L., and P. Anderson, 1986. ''Technological and Organizational Environments,'' *Administrative Science Quarterly,* 31, 439–465.

Twiss, B., 1980. *Managing Technological Innovation.* London: Longman.

von Hippel, E., 1988. *The Sources of Innovation.* New York: Oxford University Press.

Weick, K., 1979. *The Social Psychology of Organizing.* Reading, MA: Addison-Wesley.

TECHNOLOGY STRATEGY:
A GUIDE TO THE LITERATURES

Paul S. Adler

INTRODUCTION

More than a decade has passed since Richard Rosenbloom, in an essay entitled "Technological Innovation In Firms and Industries: An Assessment of the State of the Art" (1978), wrote:

> The concept of strategy offers a framework which can help us design empirical studies which might generate 'explanations' of the process of technological change in firms and industries using only a few highly salient variables. Whether this argument is correct or not is an empirical question that deserves to be answered.

The purpose of this essay is to provide a guide to the various literatures which, in contributing to our understanding of "technology strategy,"

Research in Technological Innovation, Management and Policy
Volume 4, pages 25–151

ISBN: 0-89232-798-7

might give flesh to the strategy framework for understanding technological change.

We might adapt Andrew's (1980) definition of business strategy to suggest that a technology strategy is a pattern of decisions that sets the technological goals and the principal technological means for achieving both those technological goals and the business goals of the organization. I shall tentatively define *technology* as reproducible capabilities, whether these capabilities are embodied in procedures or equipment. I shall include within the potential scope of technology strategy both the development and the deployment of technologies in products, processes, or support activities. And I shall assume that even organizations in which technology is not a critical competitive factor might benefit from a more strategic approach to the management of their technical resources and activities. It remains, however, to spell out the content of technology strategy and the organizational processes required for its formulation and implementation.

0.1 The Need for a Concept*

The development of a technology strategy framework has become increasingly urgent as firms become more conscious of the potential of technology as a competitive weapon. Friar and Horwitch (1986) list four contributors to this growing awareness: a loss of faith in other strategy doctrines, for example those based exclusively on market share; the apparent success of small high-tech firms; the priority given to technology by very successful Japanese firms; and a growing awareness of the potential contribution of manufacturing strategy and process technology to competitiveness. One might also add that technology is bound to become a more important competitive variable if technological change is accelerating. Indeed, even a constant rate of technological change implies that from one period to the next, the absolute amount of change—and the corresponding technological and managerial challenge—increases exponentially.

The result is an increasing number of firms attempting to manage technology strategically and a burgeoning case literature, but as yet no consensus on the substance of a technology strategy framework.

In part this lack of a consensus framework is due to inevitable lags in management practice. Technology strategy as an organizational process

*The heads are coded in this chapter to correspond to the numbers in the bibliography, for easier cross referencing.

presupposes a certain comfort-level with the business strategy process, and in most firms such a comfort-level has been attained only relatively recently.

But if the task of conceptualizing technology strategy is still incomplete, it is also because the research literature to date has addressed it only obliquely. In most research, the unit of analysis has tended to polarize into, on the one hand, the detailed level of discrete innovations—with a large literature on the innovation process, success and failure in innovations, diffusion of innovations, innovation transfer, etc.—and, on the other hand, the more general level of industries and national economies—giving rise to studies on the economics of R&D, international competitiveness, technology trends, industrial organization, etc. The polarization of research into these two levels leaves a gap at the firm level, limiting the value of research results for strategy analysis.

This gap is partially filled by more normative literature. But the research literature's polarization between the project level and the industry level makes it difficult to identify strategic prescriptions that are theoretically grounded or statistically tested:

- On the one hand, we need to understand how the individual innovation fits into the parent organization's stream of innovations. If, as seems plausible, a firm's technology strategy evolves under the influence of its prior innovation experience, then studies that focus on discrete innovations (such as when success factors are identified by averaging results across individual innovations) may well be methodologically flawed and lacking in pertinence (Maidique and Zirger, 1985; Radnor and Rich, 1980; Gold, 1980). Technology strategy is more than the sum of individual projects or programs.
- On the other hand, industry-level analyses may not be very useful when, as Cohen and Mowery (1985) note citing Scott (1984), the simple insertion of firm-specific dummy variables accounts for between 30% and 50% of the variance in the level of firm R&D spending. Generalizations at the national level or even the industry level may well be missing the more significant strategic levers.

To bridge this gap between discrete innovation projects and industry characteristics we need to develop a deeper understanding of how companies can develop and maintain a unity of purpose and direction in their development and deployment of technology. Moreover, by building an understanding of this organizational level of analysis, we can recast in

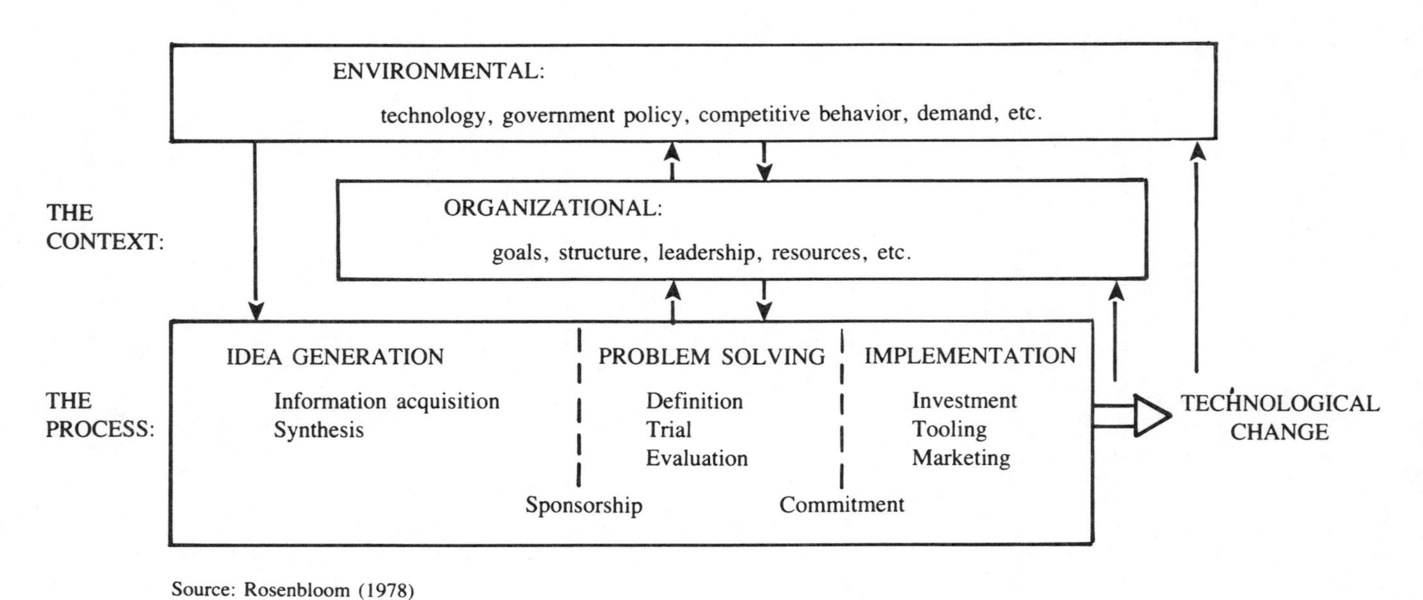

Source: Rosenbloom (1978)

Exhibit 1. Rosenbloom's Framework

more strategic form both the industry analysis—to what extent can the firm actively reshape its external environment?—and the project-level analysis—how can the sequence of projects over time be best linked to organizational objectives?

0.2 A Framework, a Model and an Outline

My discussion of technology strategy as an organization-level phenomenon takes as its starting point Rosenbloom's framework (Exhibit 1). This framework highlights the interaction of the three levels of analysis of technological change: the environment, the organization, and the discrete innovation. Following Rosenbloom, my working hypotheses are: (a) that strategy offers a framework through which we might come to better understand the "missing" intermediate, organizational level, and (b) that a strategic point of view should help us integrate these three levels, since it is centered on the organizational level and reaches out to the environment and down to the specific projects.

This paper aims to provide a guide to the literatures on these three clusters of issues. I shall proceed in turn through each of the three levels—environment (Section 1), organization (Section 2), and innovation project (Section 3). Each section will consist of a brief and selective guide to the key underlying concerns in the various relevant literatures. I have divided the bibliography into these component sections, allowing redundancy of citations where necessary.

There is first the need to understand how firms can act strategically vis-à-vis the external environment. Much has been learned about technology dynamics, the impact of societal values, the role of government policy, the evolution of demand, and the dynamics of competitive behavior. But while some studies have acknowledged that the firm has some freedom in choosing the environment—industry, market segment—in which it will compete, fewer have explored the ability of the firm to actively reshape its environment. This latter part of the problem is clearly critical for managers needing to make technological choices, for example, between attempting to break through the current performance limits of a given technology, cooperating with other organizations to create such breakthroughs, or waiting until these new technology capabilities manifest themselves "out there" in the environment. This part of my review will therefore focus on the environment viewed strategically.

The next main section addresses the literatures focused on the organization, more specifically on the nature of technology strategy decisions, and the processes by which firms come to make such decisions. In order to structure my review of the elements of technology strategy, this central part of the essay uses a heuristic device: I draw on the possible parallels between technology strategy and a simple and fairly standard model of business strategy in which internal strengths and weaknesses and external opportunities and threats inform the determination of the organization's mission, its overarching objectives, and its strategy, and this strategy is in turn operationalized as a coherent set of policies (Wheeler and Hunger, 1986).

Without prejudging the nature of the relationship between business strategy and technology strategy—this being an important issue in its own right—this framework will organize my discussion of the elements of technology strategy and their inter-relations:

- the analysis of business strengths and weaknesses finds its parallel in the identification of distinctive technological competencies;
- business opportunities and threats have parallels in strategic technology areas;
- business mission has a parallel in technology posture;
- business objectives have parallels in measurable technology objectives;
- business strategy has its parallel in a technology strategy which identifies a vector characterizing the projected development path of the organization's technology efforts and capabilities;
- and just as business strategy is translated into business policies, this technology strategy is translated into technology policies.

Since technology strategy can encompass product, process and support technologies, this discussion will assume that technology strategy is an intrinsically cross-functional matter. My characterization of technology strategy will therefore include a discussion of the role of the specific functional strategies in R,D&E, Manufacturing, Information Systems, Marketing and Human Resources. This section will also address the implementation of technology strategy in organizational structures and the technology strategy process.

If research on strategic action vis-à-vis the external environment is scanty, it seems that it is no less difficult to think strategically about the management of specific technical innovation projects. The third section addresses the literatures on project management, technology transfer, technology diffusion and new production success and failure. This research is voluminous and rich in insights; but the strategic lessons—as distinct from tactical operating recommendations—are rarely synthesized. Research has yet to clearly establish the linkages between strategy and project management. My discussion will highlight some of these links and suggest some avenues for future research.

Technology strategy as a managerial practice and as an area of research is relatively young; it is thus not well structured into a small set of competing paradigms. I shall therefore cast my net wide across a variety of disciplines that can contribute to our understanding of technology strategy, principally: organization theory, business strategy, history, economics, as well as the more pragmatically-oriented technology management literature. My aim has not been to create an exhaustive bibliography, but rather to identify the central issues and some useful entry-points into the various relevant bodies of literature. This chapter's objective is thus to offer readers a reference tool in pursuing their own investigations.

Despite my limited ambition, it is difficult not to conclude that we are far from compelling syntheses and precepts for management. But this guide will indicate how far we have come and suggest some opportunities for future work.

1. ENVIRONMENT

The heading ''environment'' encompasses technology itself, societal values, government policy, market demand, and competitive behavior. From the strategic point of view that I have adopted, the key issue underlying research on the environmental context is the possibility of proactive strategic firm behavior. Three ''levels'' of proactiveness can be identified:

- A minimal form of strategy consists of the simple recognition of necessity. Understanding the environmental pressures and trends relative to technologies, values, government, demand, and competition is a prerequisite for any strategy. Literature on these issues is abundant.
- A higher level of strategic development would consist of a framework for deciding in which subregions of these environments the firm might optimally position itself. Here a sizeable literature focuses on matching existent strengths and weaknesses of the orga-

nization with exogenous threats and opportunities presented by the environment.

- But an even higher level of strategic development would acknowledge that the firm may have the possibility of shaping the environment and creating new opportunities: advancing the technology, influencing values and government policies, changing market demand, and altering the terms of competitive rivalry. Here, it must be confessed, the literatures reviewed in this section are still weak.

1.1 Technology Dynamics

In the first part of this section's bibliography (1.1), I have included some of the more thought-provoking analyses of the underlying dynamics of technology itself. Most of this literature can be classified as aimed either at uncovering regularities or at understanding the breakdown of regularities.

To use the gestalt psychology image, most of the research prior to the 1980s saw the regularities as the "figure" and the exceptions as the "ground." At a very aggregate level, Kondratief's (1978, originally 1935) and Schumpeter's (1949) work on long waves in technological change has been renewed (Freeman, 1983; Freeman, Clark and Soete, 1982; Forrester, 1979; Wilkinson, 1983; Van Duijin, 1977; Mandel, 1981; Mensch, 1979). At a more industry-specific level, Utterback and Abernathy (1975), Abernathy and Townsend (1975), Foster (1986), Roussel (1984), and others have developed the concept of technology life-cycles.

The strategic significance of this research is of the minimalist kind: a sober recognition of the necessity of declining technology investment payoffs as technologies mature, and an awareness of the resultant changes in competitive focii. The empirical strategy research based on the notion of the product life cycle (for example, Anderson and Zeithaml, 1984; see also Rink and Swan, 1979) provides an indication of the type of results gleaned from this perspective. A large literature on technological forecasting translates this focus on regularities into management technique. A separate sub-section of the bibliography (1.1.1) identifies some useful readings in the technology forecasting area.

Assuming the empirical validity of these regularities still leaves a crucial strategic issue unresolved: as statistical constructs, their utility is much stronger retrospectively than prospectively. Knowing that a technology will eventually exhaust its potential tells the manager-strategist little about when research roadblocks that might turn out to be temporary, aren't. And even if the manager feels convinced of the relative exhaustion of a given technology trajectory, nothing tells him or her which of the candidates's new technologies will in fact "take off," nor when the new technology's S-curve will take off—these data are critical to the managerial decision of where and when to redeploy technology resources. The problem is compounded by the danger of self-fulfilling prophecies. Through a strategy of resignation, the firm can renounce any effort to influence a technology it deems mature, and if all the industry players adopt the same posture, the technology will indeed stagnate. New players, however, might disrupt the status quo, with consequences such as those experienced by the automobile industry.

Increasingly concerned with issues such as these, recent research has reversed the figure (regularities) and the ground (chaos). The late William Abernathy, after having been instrumental in establishing the technology life-cycle concept, discarded that model when he reached the conclusion that if the center of gravity of the automobile industry was shifting to Japan, it was not due to an inevitable industrial maturation that would necessarily give the advantage to low-wage-cost countries. On the contrary, he argued, the U.S. manufacturers were losing market share because the pace of technological change had quickened—indeed, that it had been deliberately accelerated by the innovative drive of the Japanese players. The industry was "dematuring" (Abernathy and Clark, 1985; Abernathy, Clark, and Kantrow, 1983; Clark, 1983). Other authors have since followed suit in their critique of mechanistic models that leave too little place to proactive firm behavior (De Bresson and Lampel, 1985a and 1985b).

The new research, having acknowledged a broader range of opportunities, has, however, left us with even fewer substantive strategic lessons, apart from the generic need for flexibility (Bhide, 1986). Some research has nevertheless been conducted on the manner in which firms can adapt to such uncertainty, such as the work by Gold and his associates (Gold, 1975, 1977); and Ansoff (1984) offers insightful discussion of the strategy process appropriate in this more turbulent context.

An important opportunity for future research thus appears to be the characterization of different types of technology dynamics in a manner that can be more directly linked to the firm's decision parameters. At the more microlevel of analysis, two particularly promising lines of research have been initiated based on reconceptualizations of technology itself. The first such approach is that based on the notion of technology hierarchies (Clark 1985; Durand and Gonard 1986), a notion that can be traced back to Alexander's (1964) design hierarchy concept. A technology can typically be broken down into a hierarchical, tree structure of constituent technology choices that relate to each other either as alternative ways of reaching similar functional ends (alternative branches) or as sub- (or super-) categories within a given set of consistent technology choices. The hierarchy concept serves to highlight the difference between innovations that reinforce existing competencies—by moving down the hierarchy to refine current technology branches—and innovations that obsolete existing competencies—by moving back up the hierarchy in order to then move down a previously unexplored branch. So far, application of this concept seems to have been more successful in generating aggregate description than detailed prescription (Tushman and Anderson, 1986).

A second promising micro line of research is based on a conceptualization of technology as involving both explicit and "tacit" dimensions. This builds on Polanyi's (1967) discussion of tacit forms of knowledge, forms in which understanding is personal and localized, rather than uniform and publicly available, and uncodified rather than explicitly codified. Explicit and tacit forms of knowledge are typically complementary in both the development (Freeman, 1986; Pavitt, 1984) and the use of new technology (Kusterer, 1978; Manwaring and Wood, 1985).

This duality gives technological development a distinctive character. On the one hand, explicit technology often has the attributes of a "public good"— neither is it used up by being used nor it is possible or desirable to enforce property rights. The relative nonappropriability of technology's explicit component (Arrow, 1962) means that market-based economies will tend to underinvest in R&D, especially in basic research. As a result, the question of public support for R&D is continually on the public agenda. Moreover, as I shall indicate in Section 2.3.2., this nonappropriability means that the problem of transfer prices in technology-intensive multidivisional firms can have no satisfactory purely economic solution.

On the other hand, the tacit component means that technological development does not take the form of a frontier of knowledge advancing at an even pace across all possible product, product and process options; instead, innovations follow distinctive and larger irreversible technological trajectories (Dosi, 1988; see also: Nelson and Winter, 1977; Sahal, 1981a, 1981b; Dosi, 1982; David, 1985; Arthur, 1985). Technologies are thus structured into paradigms, that is as exemplars (dominant designs) and heuristics (problem-solving routines). Innovation activities are strongly selective, finalized in specific directions and cumulative in the way they build problem-solving capabilities (see Kuhn, 1970, on scientific paradigms, and Dosi, 1982, 1984, 1988, on the extension of Kuhn's concept to technology).

The complementarity of explicit and tacit knowledge explains why firms do some of their own R&D rather than relying exclusively on technology vendors (Mowery, 1983). Further, it explains why a firm cannot simply decide on its technology development priorities by matching the universe of publicly-available knowledge to perceived market opportunities; it must factor into its decision the adequacy of its own "distinctive technological competencies," its "core technologies" (Teece, 1986; Pavitt, Robson and Townsend, 1987; Ketteringham and White, 1984).

These observations on the nature of technology ground another promising stream of research at a more macro level, attempting to occupy the middle-ground between generic life-cycles and unpredictable transilience. The theme of this research is that useful, albeit more modest, generalizations might emerge if one could identify different types of technologies. Pavitt (1984) identifies several distinct categories of industries based on their paths of technological change: supplier-dominated, scale-intensive, specialized suppliers, science-based. These groups face different technological opportunity spaces, different appropriability opportunities and different degrees of threat from new entrants (Pavitt, Robson and Townsend, 1987; see also Kodama, 1986a, 1986b; Dosi, 1982). In a similar vein, Levin et al. (1987) distinguish different appropriability conditions across industry clusters. Using simulation rather than empirical data, Nelson and Winter (1982) distinguish the dynamics of science-based industries and experience-based industries. Further research along these lines is needed to identify the regularities that are most germane to the individual firm.

1.2 Societal Values

Kelly and Kranzberg (1978) review the literature on the impact of society's value constellation on the legitimacy of innovation and on the particular institutions and processes involved in the innovation process. To summarize the main points:

- More advanced industrial societies have a bias towards innovations, as compared to precapitalist societies' bias against innovation (see also Rosenberg and Birdzell, 1986).

- Specific value-sets may be particularly hospitable to innovation in general—as suggested by Weber's (1930) analysis of the Protestant ethic.
- Values can affect innovation indirectly: values that encourage education, for example, indirectly encourage the emergence of innovators and raise the receptivity of users to innovation (Rogers and Shoemaker, 1971).

More specifically, values can affect the rate and perhaps the direction of technological change through their influence on the perceived desirability of specific changes (for example, the shift to public transportation advocated by some as a response to reduced oil availability in the 1970s) and the willingness to accept certain risk levels (for example, in pollutants, food additives, drugs, dams, etc.).

More detailed historical research such as Hounshell (1984) or Smith (1977) on manufacturing technology highlights the close interaction of technological opportunity and social forces in shaping technological trajectories. The articles in MacKenzie and Wajcman (1985) sample some of the research that argues that the social forces play the decisive role in this interaction. In this vein, Noble (1984) highlights social class factors at work in the development of machine-tool automation, while researchers such as Rothschild (1982), Cockburn (1983) and Harding (1986) highlight gender factors in science and engineering. New trends in the sociology of science and technology—often social interactionist in orientation—have also contributed to focusing attention on this interplay (for example, Knorr-Cetina and Mulkay, 1983; Latour and Woolgar, 1979), albeit often at the more microlevel of specific laboratories or technologies.

Apart from the impact of societal values on technology development choices, researchers have also addressed their impact on the way process technologies are implemented. Writers such as Braverman (1974) and Howard (1985) highlight the influence of labor/management conflict on job design policies for new factory and office technologies. The precise interplay of technological, economic and social factors in the evolution of job design has been the object of a burgeoning literature (see for example, Wood, 1982; Knights, Willmott and Collinson, 1985). It remains an open question as to whether the social factors outweigh efficiency considerations especially in the long run for broader aggregates of firms (Adler, 1987).

These various value effects can present themselves to the firm through the pressure of the external environment or through managers' and employees' attitudes and expectations; both may be amenable to some sort of strategic action (see Andrews, 1980, ch. 5):

- Ansoff (1984, ch. 2.5) discusses "societal strategies" focused on social responsibility and business legitimation; but he does not focus particularly on the technological dimension of societal strategy. The need for societal strategies is often thrust on the firm by "whistle blowers" who respond to the perceived societal impact of the firm's activities (see Near and Miceli, 1987).
- Many discussions of technology implementation approaches—whether for manufacturing and clerical automation (Adler, 1986) or for professional automation such as CAD/CAE (Fleischer et al., 1987)—have identified the importance of evolving employee expectations as regards their involvement in design and utilization policies (see also bibliography section 2.1.4. 7b).

1.3 Government Policy

In the third part of this section's bibliography, I have included a selection of research potentially relevant to the question of strategic behavior vis-à-vis government's role in technology. The materials on the role of government in R&D funding, patents, industrial and technology policy, regulation and education all highlight the fact that firms are not monads adrift in an ether of pure and perfect competition:

- Federal R&D expenditures and tax credits for private sector R&D (bibliography section 1.3.1) play an important role in shaping both the availability of some new technologies and private-sector research agendas. While some research addresses the interaction of private business and public agencies in the formulation of public policy (for example, Adams, 1981), we lack a general characterization of how such interaction enters into the firm's technology strategy. Case studies provide a useful starting point, especially those on the defense industries (for example, Tyler, 1986; Harvard Business School, 1986; see also Roberts, 1968).
- Patent law has a variable, but in some industries considerable, effect on innovation (1.3.2). Much research effort has been expended in analyzing the patent activity of firms distinguished by size and industry; but the literature on the patenting strategies of different

firms within the same size-class and industry remains anecdotal and slim.

- The industrial and technology policy literature is sampled in materials listed under 1.3.3. This literature provides a potentially rich source of propositions on how firms are both influenced by and influence the public policy environment. One of the more interesting issues here concerns the industrial policy implicit in government defense procurement priorities.
- Regulation of technology-intensive products and processes is another critical dimension of government's impact on a firm's technology strategy. The readings listed under 1.3.4 offer some entry-points into this research: discussion ranges from the value of specific risk-assessment techniques to the regulatory process and cooptation.
- Government's role in organizing and funding the education system clearly plays a central role in shaping the environment and resources for technological innovation. Useful material on the training of both technology developers and technology users can be found in the NSB reports listed under 1.3.5.

1.4 Demand

Thinking strategically about demand is critical to a viable technology strategy approach: what factors govern the receptivity of the market to the introduction of a new technology? The available materials are surprisingly limited.

The marketing strategy literature is, of course, the essential starting point. There is an extensive marketing literature focused on modelling the diffusion of new products (Massy, Montgomery and Morrison, 1970). But these models take technology and product characteristics as given.

The vast sociological and economic literature on the diffusion of innovations (see Rogers, 1983, and bibliography 3.1.4) could also, in principle, be a rich source of insights. But due to the focus of this research on discrete innovations and individual adopters, it remains unclear what strategic framework the sellers of new technologies might synthesize from these studies.

The literature on new product success and failure (see bibliography 3.1.2) often finds that successful innovations are pulled by demand and are based on a solid awareness of market needs. But closer analysis (Mowery and Rosenberg, 1982) shows this precept to be weak on substance and of little strategic value: a posteriori, successful products always appear to satisfy a market need; but a priori certainty on the existence and nature of that need is rare indeed.

More useful for the technology strategist is the research that helps us understand the structure and evolution of consumer needs (Urban and Hauser, 1980). Important strategic consequences follow from the research that has highlighted the interdependence of technology developers and users (von Hippel, 1977, 1978). But we need conceptual models that build on this insight. Clark's design hierarchy model (1985) of how firms learn about the nature of demand for new products provides one useful starting point. Rosenbloom and Cusumano (1987) provide another with their study of the long period of experimentation needed to identify the nature of the consumer's requirements in the VCR case.

1.5 Competitive Behavior

In the research on competitive behavior, we would expect to find a wealth of technology strategy materials. Unfortunately, the dominant economic paradigm privileges mathematical simplicity over realism when the latter threatens analytic tractability:

- Models of competitive behavior have great difficulty incorporating realistic assumptions about rivalry and technological uncertainty. The bibliography under 1.5.1 lists some of the more useful starting points. Porter's (1980) is perhaps the most popular and powerful approach. Porter's economic perspective offers particularly high leverage in the analysis of the advantages and disadvantages of being a first-mover as well as the pros and cons of technology licensing. One weakness stands out, however, in the context of our interest in proactive strategy—his assumption, taken from industrial organization theory, that competitive advantage is based on some distortion of pure and perfect competition (since perfect competition drives profits to zero). This premise, combined with the comparative-statics methodology prevalent in economics, leads Porter to bias his recommendations towards building "barriers" and protecting them (by, for example, raising customers' switching costs or holding patents), rather than focusing on how a firm can sustain competitiveness through internal efforts that enable it to move fast enough to continually recreate small, temporary barriers—if barriers be the right word. Porter's more recent work (1983, 1985) offers more insight into these issues, through a greater focus on the internal reality of the firm, in particular on its evolving technology options, but this comes at the expense of the elegance and parsimony of the industrial organization economics framework.

- International comparisons (1.5.2) provide not only an analysis of a critical environmental variable but also an opportunity to add some organizational depth to economic models. The international management literature has begun to build strategic models that allow firms a more active role in shaping their (international) environment.
- Perhaps the largest area of technology research by economists relates to the reciprocal impacts of technology and industry structure (1.5.3). The so-called "Schumpeterian hypothesis"—the proposition that larger firms have an advantage over smaller ones in being better able to routinize and/or capitalize on innovation—has been a continuing source of inspiration. Nelson and Winter (1982) have opened up important new avenues for research with their "evolutionary" theory. Overall, however, the lack of attention to organizational processes makes this literature difficult to draw into a strategic mode of reflection.
- A promising avenue of research, in which firms do actively and tangibly shape their environment, is that focused on supplier relations and vertical integration (1.5.4.). Transaction-cost analysis and agency theory allow economists at least some access to the organizational issues so central to the management of these relations. The quasi-integration of firms via repeated or long-term contracts is an important avenue of strategic interaction with the environment—I will address it more specifically in Section 2.1.4 and its discussion of cooperative policies.
- The emerging literature on the economic analysis of standards promises new insights (1.5.5). This research goes well beyond the question of mandated standards, to explore the emergence and change of de facto standards. Its focus on network externalities adds a critical element of realism to economic models: mathematical tractability may suffer, but these models highlight the path-dependence of economic processes and thus lay the groundwork for the economic modeling of the efficacy of strategic action in a world in which history really does matter.

1.6 Strategic Lessons

The research on the environment is thus rich in general characterizations, but problematic in its significance for the strategist. The key difficulty with the dominant economic approach appears to be the intractable mathematics of dynamic disequilibrium analysis in plausibly complex settings. Apart from Nelson and Winter's innovative use of simulation methods, the principal alternative to formal modelling has been the statistical analysis of larger aggregates of firms conducted in the hope of capturing the essential constraints on and opportunities for effective strategic choice. The inductive statistical approach, however, suffers from the relative paucity of combined cross-section and time-series panel data and from the nagging suspicion that it is the outlier, rather than the average firm, that provides the interesting strategic lessons.

More fundamentally, it is clear that the role of a given technology in the external environment depends to a significant extent on the nature of that technology—but we do not as yet have a good framework for characterizing and typologizing technologies. The research reviewed in Section 1.1 offers some important opportunities for the development of new conceptual frameworks, frameworks that will in turn enrich our understanding of the other facets of macro- and industry-level dynamics.

2. ORGANIZATION

In this section on the organization level of analysis, I first shall discuss the literature on technology strategy content and the contribution of functional strategies to overall technology strategy (Section 2.1); the following sub-section addresses the question of strategy implementation and in particular the role of organization structure (Section 2.2); finally, we turn to the technology strategy process (Section 2.3). In this overview, we shall find a large number of normative models, but a paucity of empirical data supporting them.

2.1 Technology Strategy Content

2.1.1 Business Strategy and Technology Strategy

A concern immediately raised by many managers relative to the notion of technology strategy is that only certain firms in certain industries should be guided by a technology strategy—by which they assume is meant a technology-dominated strategy. Underlying this assumption is often the fear that business strategy formulation might be turned over to technologists lacking market understanding. A traditional model of strategy is thus often reasserted: general management sets business goals, and managers of technical and other functions implement that strategy by orchestrating the means to meet these business goals.

Three trends in contemporary management thought and practice undermine this traditional model. First, it is being increasingly recognized that even if the functions' role is merely to implement a pre-specified business strategy, the complexity of that task warrants the articulation of multi-facetted functional strategies. Such functional strategies can help ensure that the myriad daily decisions made in each function are consistent with the business strategy.

Second, a growing body of literature focuses on the role of functional managers—and the evolution of the "means" that are their primary responsibility—in helping to set business goals. Hayes and Wheelwright (1984) argue that in more effective organizations, the manufacturing function will not only implement the business strategy, but will also shape that strategy through a proactive development of new manufacturing capabilities. Extending this proposition, Hayes (1985) argues that goal-setting too often precedes a clear understanding of the organization's capabilities. He suggests that when goals are specified in terms of desired and quantified market and financial outcomes, an invalid assumption is being made—namely, that "the world of competition is predictable and that clear paths can be charted across it much like a highway system" (Hayes, 1985, p. 117). In the increasingly dynamic and unpredictable type of competition facing most firms and especially technology-intensive firms, strategy should be a compass heading rather than a detailed itinerary: an overall direction that leaves the organization free to exploit opportunities as they arise. An understanding of the organization's distinctive capabilities and an agreement on a long-term development path for those capabilities and for the development of new capabilities are increasingly critical success factors. In this perspective, it would be illusory to think that a business can set itself meaningful goals without previously analyzing its capabilities and charting possible development paths for these capabilities. Business goals ("ends") and technology goals ("means") are, of course, inextricably intertwined; but there is a strong argument that, to the extent that the strategy-formulation process must have a starting point, this starting point should be the analysis of the organization's capabilities and their development potential rather than a set of business goals (see also Burgelman, 1984).

The third factor disturbing the traditional hierarchical division of labor between general and functional management is the increasing technological dynamism of most industries. In a technologically dynamic environment, technology and the associated strategic issues evolve too rapidly to maintain without excessive cost the top-down hierarchical model; functional managers must think and act more strategically, and general managers must lead a multifunctional and multilevel strategy dialogue. The same conclusion follows from the increasing turbulence associated with such trends as internationalization and the resulting strategic role of managers in such functions as marketing, finance and procurement. The option proposed by Hayes is therefore becoming a necessity.

SEST-Euroconsult (1984) documents the use by several leading Japanese companies of the image of a bonsai—the carefully-pruned miniature potted tree—to symbolize the relationship they see between ends and means: the firm's roots are its ability to tap into generic technologies and source sciences; its trunk is the company's distinctive competencies; its branches are the expression of these roots and competencies in specific industries, markets and product attributes (see Exhibit 2). This is a provocative image of the asymmetrical interdependence of technological and market considerations in strategy formulation.

If functional management is needed to help formulate capabilities' growth-paths, this might imply a less hierarchical, more participative model of strategy formulation. I discuss the specific roles of different functional managers in Section 2.1.5 and the strategy process as such in Section 2.3; suffice it for now to say that in practice the new, more participative model does not seem to have diffused very far (Liberatore and Titus, 1987; Cannon, 1984; Thomas, 1984; Bitondo and Frohman, 1981).

2.1.2 Types of Technology Strategy Models

This section focuses on models of technology strategy content, as distinct from the strategy process which will be discussed in Section 2.3. We should first note, however, that the distinction between content and process is not easy to maintain under the best of circumstances. Indeed, authors such as Quinn (1986, 1988) have argued that the process is the content, since the rest is likely to be a matter of ex post rationalization or symbolic exhortation. An intermediate position is suggested by Horwitch

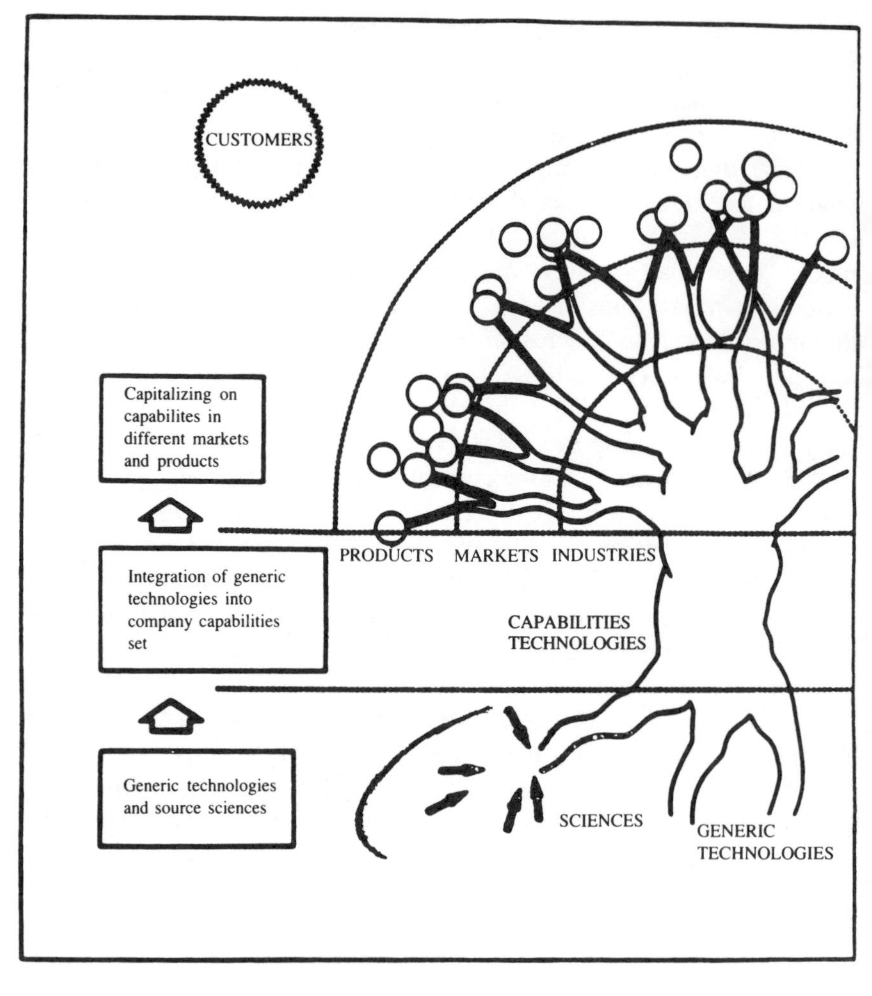

Source: adapted from SEST-Euroconsult (1984)

Exhibit 2. The Technology-Market Bonsai

and Prahalad (1976): they suggest that technology strategy—both content and process—will look quite different depending on whether the firm is small, large, or involved in a large-scale, multiorganizational program. In this section, I shall provisionally abstract from these concerns, to characterize the variety of content-oriented models found in the literature.

Content models can be differentiated along several dimensions. While clearly also a process issue, the distinction made in Section 2.1.1—between models that assume a one-way flow from business strategy goals to technology strategy goals and those allowing for a more participative goal-setting process—has content ramifications to which I shall return.

A second important dimension is whether technology is narrowly or broadly construed. Of the models referenced in the following sections' bibliography, many implicitly restrict their focus to R&D strategy or, more narrowly again, to R&D project selection. Similarly, many authors focus exclusively on product technologies, ignoring process and support technologies. Porter (1985), by contrast, makes a compelling argument that all technologies anywhere on the value chain should be included in the scope of a technology strategy, whether or not the R&D organization has direct responsibility for them.

A third dimension distinguishes between corporate-level and business-unit-level models. The former have tended to focus on technology portfolio selection and matching relative technical strength with relative business strength and with industry attractiveness (see for example, Booz-Allen and Hamilton, 1981). These technology portfolio models, like the business portfolio models on which they are modeled (Haspeslagh, 1982) suffers from several problems: (a) they offer little guidance as to how to factor into the assessment the relative opportunity for improvement of weaknesses; (b) they offer little guidance as to how to leverage positive synergies or avoid negative synergies across business units; and (c) they rely on industry-average characterizations of industry attractiveness, offering no framework for identifying the lessons suggested by particularly successful outliers. Business-unit-level models avoid these difficult issues since they presuppose a finer-grained knowledge of the given business; as a result, however, they are by nature more complex and less clearly structured.

Apart from the top-down/participative, narrow/broad, and corporate/business-unit dimensions, technology strategy models differ in their level of sophistication. By way of analogy, we might consider four levels of sophistication in financial planning, (Gluck, Kaufman, and Walleck, 1980): (1) basic financial planning, (2) forecast-based planning, (3) externally-oriented planning, and (4) strategic planning. A somewhat parallel hierarchization of general management approaches is proposed by Ansoff (1984): (1) management by control, (2) management by extrapolation, (3) management by anticipation, and (4) management by flexible/rapid response. Some technology strategy models are more planning/project selection oriented (for example, Fusfeld, 1978) and would thus rank lower on these scales, while others, such as Rosenbloom (1985), would rank higher in their focus on strategic direction-setting. Most fall in between, usually at level 2 in either analogy's hierarchy.

2.1.3 *Identifying Technology Mission-Objectives-Strategy*

To better assess these models, it is useful to refer to the business strategy model referred to in this essay's Introduction.

Before pursuing the application of this model to technology strategy, it is important to recall that despite the apparent clarity of the flow charts found in strategy textbooks, each organization needs to define its own strategy terminology and framework. Practice seems to have converged towards a model which, in some combination or other, includes the following elements: organizations assess their Strengths and Weaknesses in light of environmental Opportunities and Threats (SWOT); then, relying on and refining a shared sense of basic identity and Mission to inform the choice of a few overarching Objectives—quantifiable measures of how effectively they are fulfilling their mission—they can formulate a Strategy—a few general principles of coherence for activities that they believe will help them meet those objectives—and can then operationalize these principles in a set of Policies (MOSP).

Confronted with a problem as complex as strategy, there is no unambiguous model nor crystal-clear definitions of the key terms—each organization must define its own terminology and framework, since the result will only be as effective as the shared understanding it embodies. Moreover, despite the apparent logical flow, models such as this bear little relation to the actual formulation process. The key point, however, is that an effective strategy should encompass these elements. The accumulated practice of many firms and consultants suggests that an ends-means hierarchy of several levels is needed to manage the degree of complexity of the strategy problem as experienced all but the very smallest firms.

With these caveats, we can use the SWOT-MOSP model as a heuristic device to organize the key concepts that have been advanced as elements of technology strategy (see Exhibit 3).

Some of the more effective work to date has been focussed on the analysis of the organization's internal environment—identifying the firm's distinctive technological capabilities (DTCs)—and its external environment—identifying the relevant strategic technology areas (STAs) (see Bitondo 1986; Frohman and Bitondo, 1981; Durand, 1988; Frohman, 1978, 1980, 1985; Mitchell, 1986). DTCs and STAs can only be identified iteratively since they both inform and depend on the selected strategy. But even early iterations are immensely important, since they help develop a common language between the technology people and the other people in the organization.

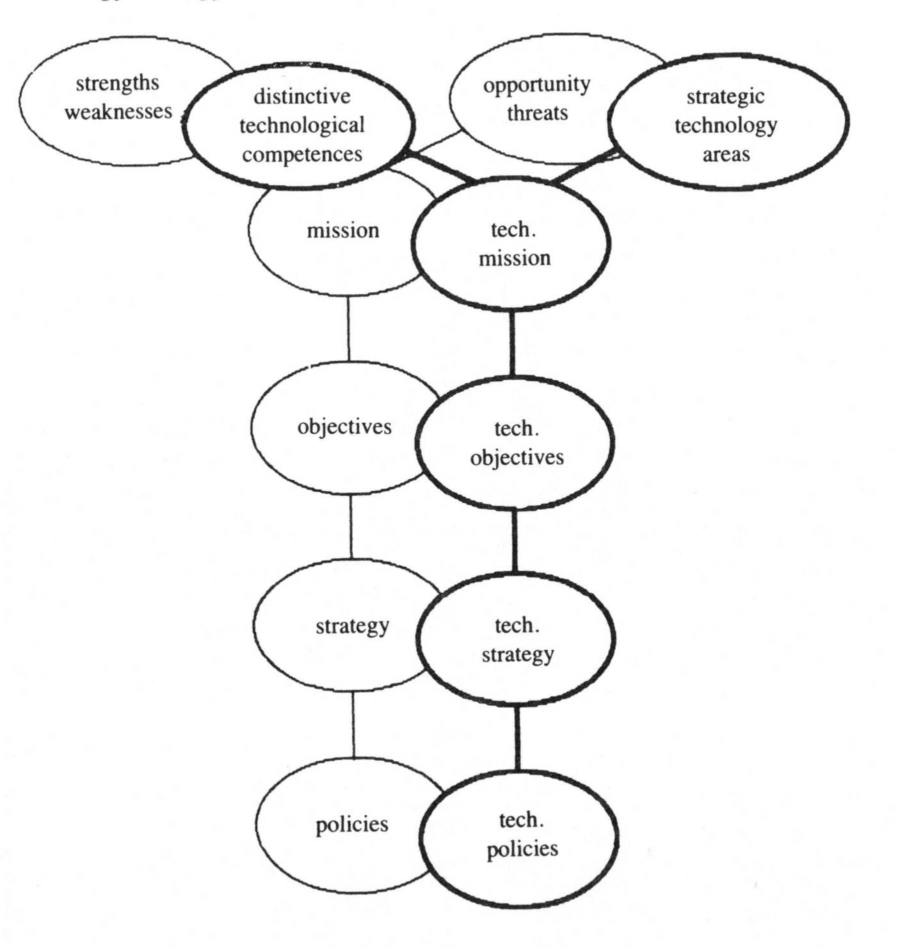

Exhibit 3. Parallels Between Technology Strategy and Business Strategy

Moreover, this preliminary process helps management to identify technologically- and managerially-relevant levels of aggregation for the characterization of its myriad specific technologies. By forcing the organization to characterize its technology issues in terms of these more generic technology domains, this step is the critical precondition enabling strategic technology management to move beyond detailed project selection; the characterization of DTCs and STAs empowers senior managers by giving them a strategic handle on the broad contours of the organization's technology activities. If, further, the organization assesses each of these technologies as base, key, pacing, (A.D. Little, 1981) or emerging

(Booz-Allen and Hamilton, 1981), it will have already created for itself a knowledge-base far superior to that of its typical competitor.

The confrontation of these technological strengths, weaknesses, opportunities and threats with the corresponding business situation lays the foundation for defining technology mission, objectives and strategy.

Technology has been recognized as a factor shaping business mission (Abell 1980); but within a technology strategy we can also define a specifically technological mission. Some of the associated issues have been addressed in discussions of technology posture. Much of the literature on posture follows and extends the classic Ansoff and Stewart (1967) distinction, elaborated by Maidique and Patch (1980), between first-to-market, follow-the-leader, applications engineering, and me-too/low-cost postures. More recently other typologies have been advanced. In Exhibit 4, I have compared the principal typologies.

The first difficulty with most of these posture frameworks is their focus on product technology: a low-cost orientation associated with a me-too product posture can sometimes imply a very aggressive, inventive, leadership posture in process technologies.

Second, many businesses find themselves adopting different postures for different elements of their technology activities. The previous paragraph indicates one possible divergence between product and process technology postures; moreover, in any but the simplest businesses, the firm will adopt different postures towards specific subsets of product and process technologies (Goldring, 1979).

Third, it is not clear whether these posture options offer sufficient coverage of the diversity of industry settings. They tend to take as their reference point high-technology industries. This limitation is increasingly serious as an increasing number of traditionally less technology-intensive industries confront more rapid technological change.

Whatever the list of posture options, posture analysis, like portfolio analysis, and despite its popularity, must be handled with caution (Ansoff, 1984):

- It is often a merely ex post result. To take one example: not many—perhaps too few—companies choose to be fast-followers; many more try for first, but just end up coming in late.
- The choice of an appropriate posture is usually seen as reflecting given strengths and weaknesses and given external opportunities and threats; this approach risks underestimating the potential of

Ansoff and Stewart (1967)	*Maidique and Patch (1980)*	*Twiss (1980)*	*A D Little (1981)*	*Bitondo and Frohman (1981)*	*Martin (1984)*	*Freeman (1986)*
First-to-market	First-to-market	Offensive market creation	Technology leader	Technology inventor	First-to-market	Offensive
Follow-the-leader	Second-to-market		Technology follower.	Technology innovator	Follow the leader; defensive	Defensive
Application engineering	Market segmentation	product improvement	Technology niche development	Technology application	Applications engineering; interstitial	Opportunist
		Defensive				
Me-too; low-cost	Late-to-market; cost minimization	process improvement	Technology rationalization	Technology avoider	Imitative, me-too	Imitative
		Licensing, absorptive	Joint-venture		Absorbant	
					Branch-plant (supplier)	Dependent
					Harvest	Traditional

Exhibit 4. Proposed Posture Typologies

proactive strategies that seek to transform the organization's current capabilities and to change the technology or market parameters.

- Third, posture analysis' attractiveness seems above all to be at the corporate level, where concern is focused on the balance of corporation's portfolio; but this begs the question raised in the preceding point, and care must be taken to plan insofar as possible for both positive and negative synergies across businesses.

On the one hand, these difficulties might justify Porter's (1985) approach: he identifies only two technology postures—leadership and followership—and then explores the ramifications of both postures when associated with the various generic business strategies. Perhaps researchers have been trying to make posture carry too great a conceptual weight in the characterization of technology strategies.

On the other hand, a key value of posture analysis emerges when we highlight the parallels between posture and the "mission" element of the standard business strategy model: the discussion of technology mission/posture should clarify the "identity," "ethos," and "vision" of the technologists of the organization. In determining the business mission, the link between strategy and organizational culture is critical (Pascale and Athos, 1981; Schein, 1984); this link is no less critical for technology strategy, as is clear in Burns and Stalker's (1961) discussion of the differences between organic and mechanistic R&D contexts or in the research that has discussed the overall cultural contexts most likely to foster innovation (for example, Quinn, 1985; Peters and Waterman, 1982; Maidique and Hayes, 1984). Not only the development, but also the implementation of technology is often informed by an ethos: Walton's (1985) analysis of automation and industrial relations highlights the contrast between "control" and "commitment" philosophies of workforce management in the posture that firms take towards implementing process technologies.

Moreover, mission is defined not only by these value-orientation issues; it can also be defined in more substantive terms. Porter (1985) proposes using the value chain to identify potentially critical points of technological leverage, and shows how the product technology and process technology mission should differ across his four generic strategies. Rosenbloom (1985) suggests that by distinguishing domains of use and domains of technology, the technology mission of an R&D organization can be classified as product-targeted (one use/one technology), application-targeted (one use/many technologies), technically-targeted (many uses/one technology), or exploratory (many uses/many technologies).

Case studies such as Chaparral Steel (1987) show how an organization can derive significant competitive advantage from a coherent ethos of technology innovation—in this case, an ethos one might call a process-focussed application-targeted "aggressive incrementalism"—that is consistent both with the spectrum of technological opportunities they face and with their business strategy.

The next element of technology strategy—technology objectives—is not well documented. The essential function of "objectives" in business strategy is to provide an objective, measurable yardstick against which the organization can calibrate the extent to which it is fulfilling its mission. What are the measurable objectives that a technology strategy could use? Chaparral Steel has a very clear objective: hours per ton of steel. In other contexts, it may be interesting to track the percentage of sales accounted for by products introduced within the previous three years, as suggested by the PIMS questionnaire. As more companies articulate technology strategies, we shall develop a clearer understanding of the range of options for identifying objectives.

The next element of technology strategy is a guiding principle for getting from here to there—strategy per se. If technology mission defines a posture, technology strategy should identify the overall direction of movement for the organization's technological resources and activities. But the formulation of technological direction still lacks a consensus framework. Several dimensions have been discussed in the literature.

- The mix of basic vs applied vs development is one breakdown commonly used (Maidique and Patch, 1980; Ansoff and Stewart, 1967; Galbraith and Kazanjian, 1983; Burgelman et al., 1988; Cooper, 1984; Foster 1986). Clearly, however, it is insufficient, since the organization needs to position itself at different places along this spectrum for its different technologies.
- A second key dimension, one that seems less subject to controversy, is that suggested by Abernathy (1978): product versus process focus. Abernathy provides a rich analysis of how Ford's early strategic focus on process refinement and cost-efficiency came to impede technology development efforts oriented towards project innovation. This distinction seems robust at least for fabrication assembly operations, even if the strategic rule of thumb that some base on it are controversial (cf. discussion of process innovations displacing

product innovations in the technology life-cycle discussed in Section 1.1).

- The distinction between incremental and radical innovation focii is often considered fundamental (for example, Normann, 1971; Ettlie, Bridges and O'Keefe, 1984). Johnson and Jones (1957), Rumelt (1974) and Meyer and Roberts (1986) suggest extending this to create a 2-dimensional matrix of technology newness and market newness. Branching out into new technological and market domains involves a different set of challenges from refining the current technical and marketing capabilities; both risks and rewards can be greater. Firms thus face a strategic choice in how active they want to be in the pursuit of emerging and pacing technologies and in their thrust into new markets. Two comments on these types of distinctions are in order. First, examples such as the video recording industry (Rosenbloom and Cusumano, 1987) suggest that the incremental/radical classification scheme may sometimes be too crude to capture the distinctiveness of many modern innovations that combine numerous incremental changes to form a genuine breakthrough. Second, the radical/incremental distinction assumes that the greater the degree of technical newness, the lower the probability of technical success. But there may be important differences in both the degree and the type of uncertainty created by different types of newness. Reinforcing versus obsoleting changes of comparable magnitude might create very different challenges, as suggested by the hierarchy model discussed in Section 1.1. (Tushman and Anderson, 1986, provide some evidence for this proposition; see also Pierce and Rosegger 1970, and Mansfield, 1961, on the importance of how much of the firm's equipment base is obsoleted by an innovation.) And there is often a big difference between attacking "new to the world" and "new to the company" domains.
- Another commonly-made distinction is between market-pull and technology-push innovation. But as suggested earlier (Section 1.4), and as shown by Mowery and Rosenberg (1982) in their critical review of the literature purporting to show the dominance of demand-pull innovations, "both the underlying, evolving knowledge-base of science and technology, as well as the structure of demand, play central roles in innovation in an interactive fashion, and neglect of either is bound to lead to faulty conclusions and policies" (p. 195) (see also Kiel, 1984). Technology managers might find

something of value in the market-pull vs technology-push distinction in characterizing their mission (posture or ethos); but it is probably not a robust guide to strategic technology direction-setting. Finally, the question of compatibility and standards has, over recent decades, become a strategic concern for broader segments of industry. Research at the industry level of analysis (1.1.5) has progressed much further than research on the strategic options available to the firm (Gabel, 1987; Hariharan and Prahalad, 1988). But it is clear that in some industries a strategic orientation vis-à-vis compatibility is a key success factor.

The technology strategy statement should identify the major criteria for resource allocation. Mitchell and Hamilton (1988) suggest distinguishing funding approaches for three different types of activity: knowledge-building (to be funded by cost allocation), strategic positioning (funded using an options valuation) and business investment (funded on ROI).

The availability of these broad categories should not, however, obscure the key difficulty: the organization needs a substantive characterization of its technological priorities. Meyer and Roberts (1988) present some provocative evidence on the competitive value of a clear focus for the firm's core technologies. But as already noted, identifying these core technologies is a nontrivial task for which we do not as yet have robust frameworks.

Further empirical research on all these elements of technology strategy is needed. In-depth case studies have proven fertile; but there are not many of them available. Even if the pressure to articulate a comprehensive technology strategy is growing, and even if the maturation of the business strategy process in U.S. industry seems to make the formulation of technology strategies a logical next step, relatively few organizations have as yet confronted the challenge in a systematic manner. Statistical analysis can perhaps help in identifying the key issues that such strategies must confront; but researchers cannot get too far ahead of practice in a domain as unstructured as this, and they are hampered by the absence of data-bases with relevant variables. The PIMS data-base is perhaps the only larger data-base that includes some of the variables relevant to these concerns (for example, Clark and Griliches, 1984; Clark and Hayes, 1985; Horwitch and Thietart, 1987; on PIMS see Anderson and Paine, 1978; see Meyer and Roberts, 1986, for an example of the strengths and weaknesses of smaller data sets.)

2.1.4 Operationalizing Technology Strategy: Policies

Mission, objectives and strategy state the principles and operating policies operationalize these principles. The organization thus needs (a) to identify the policy domains relevant to technology strategy, and (b) to make a consistent set of policy choices in those domains.

Unfortunately, the bulk of the literature on technology strategy frameworks has remained at the mission-objectives-strategy level of analysis and has rarely addressed the policy level in a systematic manner. This weakness is of considerable importance if, as the classical strategy framework suggests, policies are the operationalized substance of strategy. It is intriguing that some strategy models skip policies to go directly from strategy to specific projects. Policies and procedures have perhaps become the symbols of rigid bureaucracy. But the Japanese success in using policies such as "Just-in-Time" as tools for organizational learning reminds us that policies are an important means for building unity of purpose.

The bibliography under 2.1.4. lists some of the materials that have served as a guide to my effort to identify the key technology policy domains.

It is useful to compare efforts to identify technology policy domains with the policy categories proposed by Hayes and Wheelwright (1984) for manufacturing strategy: capacity, facilities, process technology, vertical integration, workforce, quality control, production planning and inventory control, organization. Ansoff and Stewart (1967), by contrast, suggested five key policy domains in their characterization of "technology profiles": research versus development mix, extent of interfunctional downstream coupling, length of product life-cycle, R&D investment ratio, and proximity to the state-of-the-art. Clearly the policy domains relevant for one function may not be relevant in another.

This comparison prompts two additional general observations. First, each domain should represent a (set of) choice variable(s) for the organization. Ansoff and Stewart's inclusion of length of the product life-cycle thus seems incongruous, since this is an industry characteristic, not a choice variable at least in their analysis.

A second lesson from this comparison is that there is probably not any theoretical way to generate the "right" list. The identification of policy domains needs to be adapted for each organization, and it will only be through the accumulation of experience by managers trying to make sense of their enormously complex strategy problems that we will be able to cull the most broadly useful set of candidate policy domains.

In Exhibit 5, I have identified the principal policy domains that have been discussed in the literature to date. There is no unanimity on how to cluster these categories; Exhibit 5 suggests as a starting point a breakdown into policies regarding technical capabilities and technical processes. Capabilities capture a "state" view of technology: technical workforce, the structure of technical organizations, the level of funding, equipment in technical functions, and technology sourcing. Processes capture the way those capabilities are set into motion: project selection, management of technical projects and ongoing technical operations, interfunctional and interdivisional relations. Separate sections in the bibliography offer some entry points into the literature on each of the policy domains and some of their sub-headings.

1. TECHNICAL PERSONNEL

a. Skills
  - training and development
  - hiring criteria
  - technical vs. managerial skills and careers
  - tracking skill mix

b. Evaluation and rewards
  - performance measurement
  - pay systems
  - pay level vs. industry
  - incentives for entrepreneurial behavior

2. ORGANIZATION OF TECHNICAL RESOURCES

- functional vs. matrix vs. project
- degree of centralization, formalization, specialization
- facilities location and design for communication

3. FUNDING

- level funding compared to competition
- sensitivity of funding to corporate earnings
- different criteria for R vs. D vs. E (or other categorizations)
- funding central R&D through internal corporate tax vs. internal contract vs. external funding
- timing of additions and reductions in R,D&E personnel levels

Exhibit 5. Technology Management Policies

4. EQUIPMENT IN R,D&E

- [all the other policies should also apply to the technologies, organizations and personnel that support the technical staff]

5. INTERORGANIZATIONAL RELATIONS

a. Sourcing, commercialization, joint development, standards
    - internal development vs. collaboration vs. buy
    - criteria for vendor selection and renewal
    - cooperation and stability vs. arms-length market relations with vendors and customers
    - licensing
    - patents

b. External information
    - information sharing vs. secrecy
    - information collection and dissemination

6. PROJECT SELECTION AND TERMINATION

- use of formal vs. subjective criteria
- different criteria for different types of projects
- frequency and depth of senior management involvement
- fit with product line and product generation plans
- post-project audits

7. MANAGEMENT OF TECHNICAL PROJECTS

a. Development procedures
    - milestones vs. schedules in project management
    - different management and control procedures for different types of projects
    - other functions' involvement in projects (when, degree of influence, conflict resolution)
    - senior management involvement in project reviews
    - tracking projects
    - choice of project management tools

b. Implementation procedures
    - process design for implementability, flexibility, serviceability
    - training for new tools
    - employment security
    - deskilling vs. enabling principles for design and implementation
    - user participation in design
    - union involvement in technology choices

c. Staffing technical projects
    - size of project teams
    - project team longevity
    - dedicated vs. part-time staffing of project teams

Exhibit 5. (Continued)

8. MANAGEMENT OF ONGOING TECHNICAL OPERATIONS

a. Quality control in technical operations
    • cost of quality measurement
    • focus on quality of output vs. quality of process
    • documentation
b. Schedule and cost control in technical operations
    • tracking processing-times and wait-times in operations
    • capacity utilization vs. time-to-market objectives
    • senior management involvement
c. Regulatory compliance
    • liability policies
    • whistle-blower safeguards
    • line vs. staff responsibility for regulatory compliance
d. R,D&E performance measurement
    • choice of indicators
    • internal vs. external comparisons
    • technical excellence vs. business contribution

9. INTERFUNCTIONAL INTERFACES

a. With manufacturing
    • extent of early downstream involvement in design
    • informality and continuity of relationship
    • product design for manufacturability
b. With marketing
    • technical strength of marketing staff
    • market contact for technologists
    • involvement in design projects

10. INTERDIVISIONAL RELATIONS

a. Sharing resources
    • division managers' compensation structure and evaluation criteria
    • structures for joint initiatives
    • forums and processes for information sharing
    • transfer pricing
b. Formation of new lines of business
    • mechanisms for funding unplanned initiatives
    • internal vs. external location of new ventures
    • degree of control over new ventures
c. Relations between corporate R&D and divisions
    • personnel rotation between divisional technical staff and central R&D
    • funding projects
    • technology transfer procedures

Exhibit 5. (Continued)

Several comments on this table are necessary.

First, I have excluded some policy domains from some authors' lists when I thought they were:

a. more properly industry characteristics than choice variables (for example, product life-cycle);
b. more properly strengths/weaknesses of the organization; or
c. more properly elements of the mission-objectives-strategy set (for example, relatedness of technologies, technological focii.)

Second, there is clearly some overlap between these domains. My criterion in disaggregating these interwoven policy domains was the extent of overlap in the relevant literatures.

Third, it is presently not clear how we should characterize the key dimensions of choice in many of these domains. More research could usefully be devoted to better specifying not only the policy domains but also the principal policy options in each of them.

Fourth, even though I have included a separate policy domain for engineering tools (2.1.4.4), Exhibit 5 is constructed to suggest that policies need to be articulated for all three broad classes of technology—product, process and support. The relative strategic significance of each of these classes can only be determined on a case-by-case basis and by an in-depth analysis. Even if this analysis reveals the overriding role played by one class of technology, the others need to be managed strategically, since the interaction of the three classes typically presents powerful synergies—positive or negative. The complication this introduces into the strategy process will be addressed in Section 2.1.5.

Beyond these general comments, two comments on specific domains are appropriate.

First, among the policy domains I have identified in Exhibit 5, interorganizational linkages for both sourcing and commercialization of technology have become increasingly important issues (2.1.4.5). Friar and Horwitch (1986) document the dramatic turn outward in U.S. companies' technology sourcing over the last decade. This out-sourcing can take numerous forms: acquisition of other companies, licensing, and, with increasing frequency, cooperative R&D. In part, this extroversion seems to be fueled by a frustration on the part of general managers with the performance of their internal R&D labs. Another factor is probably the prevalence of diversification strategies in this period, with the attendant multiplication of the number of technological domains potentially relevant to any business. Finally, the general tendency towards an acceleration of technological change increases the value of external linkages to related source sciences and generic technologies.

Apart from the relatively tangible issues of cost, time and risk, both licensing-in and licensing-out require subtle analysis of the state of the current knowledge-base and the likelihood that licensing could inhibit or stimulate subsequent development of the organization's distinctive capabilities. Japanese industry has been particularly effective at using licensing-in from the U.S. and Europe to build its knowledge-base (Peck and Goto, 1981). And U.S. multinational firms have had to think strategically about licensing-out in the international arena (Contractor, 1981; Telesio, 1979; Wells, 1983, chapter 7).

Beyond these market-mediated relations, interest is growing in nonmarket forms of collaboration—joint research ventures, research consortia and the like. The theoretical interest of these interfirm strategies lies in the way they challenge the assumption of competition behind most models of strategy and firm behavior. These cooperative strategies are clearly embedded in a sea of competition; but it is important to understand how cooperation and competition are increasingly interwoven in modern industry (Ouchi and Bolton, 1988; Von Hippel, 1987). Our understanding of competition itself will be changed as we come to better understand the nature of cooperation. Given its potential significance, the bibliography on sourcing and commercializing includes some references to this broader question.

Second, the policies pertaining to the management and control of technical projects (2.1.4.7) and ongoing operations (2.1.4.8) need some elaboration. While there is a great deal of material on project management, relatively little has been written about the management of the underlying, ongoing technical operations. A growing stream of research shifts the focus to the system in which engineers and projects operate (see Liker and Hancock, 1986).

Three implications of this shift should be mentioned. First, the control of engineering quality is a growing preoccupation in U.S. industry; but to date almost all the literature on R,D&E quality has focused on the quality of the R,D&E output—usually a set of specifications—not on the quality of the R,D&E process (see for example, Wolff, 1976). The new aggressiveness of manufacturing managers in building quality into their processes—as opposed to inspecting it into their products—is beginning to spread to R,D&E operations, as some managers capture data and focus attention on such process issues as the proportion of designs that fail at standard test points, the average number of prototype iterations or the number of engineering changes after manufacturing release.

In a similar vein, the reduction of "designs-in-process" inventory is becoming more urgent as firms attempt to respond to design-to-market cycle-time pressure (Bower and Hout, 1988). While much has been accomplished in developing methods for scheduling individual projects, much less research has been conducted on scheduling the work-load of an entire engineering organization across multiple concurrent projects. The common assumption, if implicit, in many engineering organizations is that this macro-scheduling should attempt to maximize engineer utilization. The "Just-in-Time" manufacturing doctrine makes it obvious why this rule slows down new product development.

Finally, this shift rebounds on development project management policies. One insight to have emerged from recent research is that Crosby's (1980) "quality is free" argument for manufacturing probably applies to new product development time too. New product development time trades off against product quality or cost when "everything else is held constant" (Scherer, 1966; Mansfield et al., 1971); but across a series of projects, managers, appropriately encouraged, will typically discover ways to accelerate the development cycle by managerial and technical innovations that simultaneously improve time, cost and quality for products in a given class of performance characteristics (Clark, Chew, Fujimoto, 1987; Stalk, 1988).

There is often some resistance on the part of professional employees to the idea of active management of technical operations (in the name of individual professional autonomy) and to the idea of comparing performance across several projects (in the name of the creative idiosyncracy of each project). But most of most engineers' time is spent on routine activities. So aggressive management of process efficiency could help them work smarter and increase the ratio of creative to routine time.

2.1.5 Technology Strategy and Functional Strategies

In opening this section on technology strategy content, I raised the general question of the relationship between business strategy and technology strategy, and suggested that the role of functional managers might need to be more proactive than the conventional strategy models assume.

In practice at the business-unit level, this means that the technology strategy of the business-unit cannot be sole property of any one function: there is typically considerable potential synergy between product, process and support technologies, and there is always considerable interaction between technology choices and other functional strategy decisions (personnel, finance, etc). The overall technology strategy of the business has to be elaborated and implemented jointly by all the functions (Davis 1986).

Similarly, at the corporate level, the potential synergy between divisional technology initiatives and the importance of the linkage between central R&D and divisions (wherever there are central R&D facilities) argues for the elaboration and implementation of a corporate technology strategy though a joint process that is both inter-divisional and inter-level (division/corporate). A growing number of firms are therefore nominating Chief Technical Officers (Adler and Ferdows, 1988) to help orchestrate this process.

In other words, technology strategy is like human resource strategy, in that it should be the outcome of a dialogue between all the functional managers, business unit managers, corporate executives, as well as specialist staffs.

Given the intrinsic complexity of this hypothesized strategy process, it is not surprising that some general managers resist the idea of a technology strategy, doubting that so many interacting objectives can be simultaneously optimized in the real world of limited managerial resources.

The literature on functional strategies suggests, however, that the content of such a cluster of strategies can be appropriately simple (Hayes and Wheelwright, 1984). Moreover, in reality, functional managers already play a critical, if informal, role in the formulation of strategy (Guth and MacMillan, 1986). More explicit models of such a multifunctional multilevel strategy process thus seem the logical next step in the refinement of strategy (Bourgeois and Browdin, 1984; Newman, Logan and Hegarty, 1987). Techniques such as explicit "mapping" of products, markets and technologies can powerfully assist the interfunctional, interlevel dialogue (Hayes, Wheelwright and Clark, 1988; Willyard and McClees, 1987).

While the previous section identified interfunctional interface policies, the subsections that follow return to strategy more broadly defined, to identify some literatures that could contribute to a better understanding of the interaction between function-specific technology strategies, the overall technology strategy of the business unit, and its business strategy. The relevant bibliography in each of the functional areas is considerable; so here even more than elsewhere in this chapter, my objective is to suggest some entry points rather than provide a comprehensive listing.

In reviewing the functional areas individually, one can note a rather uneven development of functional strategy concepts. Most of the functional management literature reflects the primarily day-to-day focus of the functional manager. But the overall trend seems to be towards this new model in which functional managers have longer time-horizons and play a more strategic role.

2.1.5.1 R,D&E Functional Strategy. While Section 2.1.3 discussed the general question of the role of technology in business strategy, the bibliography under 2.1.5.1 focuses more specifically on the relationship between functional goals and business strategic goals.

This section's bibliography raises two issues concerning the scope of R,D&E functional strategy. First, as already noted, there is an unfortunate "ethnocentric" tendency to reduce technology strategy to the narrower question of R,D&E functional strategy. The effect of this R,D&E "imperialism" is often that R,D&E functional strategy pays insufficient attention to the strategic—as distinct from tactical—management of its interfaces with marketing and manufacturing. As suggested in Exhibit 5, the question of functional interfaces is too important to be left to the discretion of individual project managers. The strategic commitment of the entire management team is a precondition for the effectiveness of such interfunctional activities as market and producibility analysis for new product designs.

Second, when organizations separate corporate R&D from divisional Engineering, strategic planning tends to focus on the corporate group, with the result that one of the few business-level functions not to have advanced a more strategic view of its future is the Engineering function. Often split between product design and manufacturing engineering, the business-unit engineering function has typically remained focused on implementing the business strategy. If it formulates an explicit functional strategy of its own, it is typically only in the area of new product choices.

2.1.5.2 Technology and Manufacturing Strategy. Bibliography under 2.1.4.7b discussed implementation policies for operations technology and bibliography under 2.1.4.9a focused on the R&D/manufacturing organizational interface; but manufacturing technology itself is often a key competitive factor. Within the broad scope of manufacturing strategy—as the concept has been developed by Skinner, Hayes, Wheelwright, Buffa, and others—the strategic management of process technology clearly has a critical role to play. Process technology is both a dimension of the firm's overall technology strategy and a policy domain in the firm's manufacturing strategy. The concept of the product/process matrix (Hayes and Wheelwright, 1984), by clarifying the need to align product volume and variety with process technology automation and flexibility, provides a powerful link between manufacturing strategy and technology strategy.

As previously noted, the manufacturing function is important in another regard too, since it is perhaps here that the idea of functional strategy has been best articulated. Hayes and Wheelwright (1984) have nicely presented the logic of the functional strategy argument in the context of their elaboration of manufacturing strategy: manufacturing strategy does not have to limit itself to achieving a neutral or even an internally supportive status for manufacturing, but can actively shape the competitive strategy of the firm by the new opportunities it creates.

2.1.5.3 Information Systems Strategy. In some industries such as banking and airlines, information technology has the capacity to redefine the business. Even where its power is less visible, advances in information processing technology in its various forms make it sufficiently important to warrant strategic management. MIS departments are therefore developing a longer-term strategic sense of their contribution to competitiveness. Indeed, this development has been so dramatic as to prompt some firms—most noticeably in the banking and airline industries—to designate Chief Information Officers capable of explicating the link between data processing/telecommunications technology on the one hand and business strategy on the other.

2.1.5.4 Marketing Strategy. Marketing researchers have been very active in defining the factors of new product success. Unfortunately, most of this research, like much of the research in R&D management, has been focused on individual products. This section's bibliography highlights some of the more useful literature on the relationship between marketing strategy and technology strategy; bibliography on the organizational interfaces between marketing and R,D&E functions can be found under 2.1.4.1.9b.

2.1.5.5 Technology and Human Resources Strategy. Finally, and cutting across all the preceding functional strategies, human resources strategy has multiple intersection points with technology strategy. The bibliography 2.1.4.1 and 2.1.4.7 identifies several policy-level intersections; bibliography 2.1.5.5 focuses on the higher, strategy-level intersections.

These strategy-level issues pertain to both technology development and technology implementation. The latter part of strategic human resources planning is often particularly poorly handled, leaving line managers in a reactive posture vis-à-vis changes in headcount and skill requirements associated with the automation of shop-floor, office or technical operations. The industrial relations component of technological change is perhaps the most challenging of these human resources issues (see, for example, Hawthorne, 1978, ch. 10; Kennedy et al., 1982; Lansbury and Davis, 1984; Walton and McKersie, 1988; and other materials under 2.1.4.7b).

Further research on the technology strategy–human resources strategy connection seems particularly urgent given the emerging consensus that U.S. competitiveness has suffered less from a lag in innovative ideas and more from industry's difficulties in effectively mobilizing its human resources for the commercialization and implementation of those ideas.

2.2 Technology Strategy and Organizational Structure

Implementing technology strategy—as in any type of strategy—involves consideration of two broad sets of issues: designing the appropriate organizational structures and pursuing the appropriate projects. (My expansive policy framework has led me to include material on many of the other issues often included under implementation in the bibliography on policies.) Section 3 will take up implementation issues as they appear at the level of individual projects; in this subsection, I shall focus on organizational design for the effective implementation of technology strategy.

2.2.1 Theoretical Models of Technology-Structure Linkage

Organizational sociology has devoted considerable effort to understanding the nature of the strategic fit between organizational structure and technology.

A first stream of research focuses on process technology and organization structure (see reviews by Gerwin, 1981, and Scott, 1988). Depending on such features of the process technology as its degree of complexity (Woodward, 1965), the analyzability and the frequency of exceptions (Perrow, 1967), and the degree of workflow integration (Hickson et al., 1969), workers need more or less discretion, and the organization will therefore be structured quite differently in its formalization, centralization, span of control, degree of specialization and so forth. The research to date shows that while there are important linkages between technology and organization structure at the task and work-group levels, the relationship weakens considerably as the scope of the analysis broadens to larger units. As the scope broadens, the number of distinct technologies being used tends to increase, and variables such as strategy and the size of the organization come to play a more important role than technology and task in determining structure.

A second stream of research focuses on the organization's adaptation not to given process technologies but to changes in the technological environment. In this research, the organization is viewed as an "open system" and other organization's technological changes create new challenges for managers (Dill, 1958; Thompson and McEwen, 1958). Burns and Stalker (1961), Hage and Aiken (1969) and Lawrence and Lorsch (1967) thus analyze the organizational structures most appropriate to more and less volatile technological and task environments.

A third research stream shifts the focus from how organizations adapt to given technologies and environments to how they attempt to reshape the environment. Much of this research has focussed on how organizations attempt to protect themselves from environmental uncertainty by the use of "bridging" strategies such as bargaining and cooptation (Thompson and Bates, 1967; Pfeffer and Sahancik, 1978; Pfeffer and Leblebici, 1973; Pfeffer and Nowak, 1976).

The organizational component of efforts to change the specifically technological environment has been explored by the research program of Van de Ven and his colleagues at the University of Minnesota (Van de Ven, Angle and Poole, 1988). The richness of this team's research highlights the multiplicity of forces at work and possible levels of analysis. This complexity creates a serious challenge to the norms of theoretical simplicity and parsimony to which most organizational research has aspired.

Since Child (1972), we know that strategy mediates the technology-structure relation. But once organizational theory allows this mediation, it appears to open an apparently boundless universe of new questions which problematize the notions of technology, structure and strategy. Some of these theoretical questions have been:

- The firm reacts not to the environment or the technology as objective realities but to its perception of them—thus the question of "enacted" environments: how do firms process images of their environment? How broad is the margin of interpretation?
- The firm selects its technologies and disposes of a considerable margin of freedom in its implementation of them. One of the moti-

vations governing these choices is often their impact on the type of organization structure that key decision-makers want to see maintained or created. Clearly these organizational design choices are somewhat constrained by technical factors and economic forces, but which comes first, the chicken (technology), the egg (structure) or the intention (strategy)?

- Strategies have multiple possible motivations, and different stakeholders inside and outside the firm all have their own strategies—thus the question of whether a firm can be said to have a superordinate strategy, and if so, what conditions are required for such a commonality of purpose to arise. Moreover, there is typically a multiplicity of stakeholders external to the organization and an interdependence of groups within and outside the organization—thus the question as to whether the boundary dividing organization from environment is always the most salient one.

Research on technology strategy seems to confront organization theory with the need to move to a qualitatively higher level of theoretical sophistication and complexity.

2.2.2 Structures for Innovation

In the face of this complexity, many organizational researchers have veered away from the larger-scale issues of technology and organizational design; others, however, with a stronger interest in technology issues have pushed ahead in more specialized directions.

First, there is a considerable literature on the design of organizations to maximize the most appropriate face-to-face contacts and information flows. Here the work of the MIT group around Roberts, Allen and von Hippel has been central. These authors' research serves as an important counterweight to the sometimes excessive zeal of those more preoccupied by the design of elegant and complex algorithms for technology strategy formulation and project selection. Going back to Roberts' (1962, 1967) early criticisms of these formalist tendencies and his criticism of the U.S. Department of Defense's efforts to come to grips with its enormously complex technology management problem, the MIT research has shown that the beginning of wisdom in the technology strategy domain lies in an acute awareness of both the powerful effect exercised by the organizational context and the great weight played by the aptitudes of key individuals and by interpersonal networks. Thus the central role accorded by these authors to communication patterns within and between technology functions, between technologists and marketing, and between developers and users. These concerns are reflected in the policy bibliographies on personnel, structure and interfaces.

A second, emergent body of research addresses organization structure from a more cultural point of view. Ouchi (1980) argues that under conditions of performance ambiguity typical of technologically dynamic industries, ''clans''—organizational forms based on shared values—will outperform market relations or bureaucratic structures. The research by Pascale and Athos (1981), like that of Peters and Waterman (1982), has served to highlight the critical role of values as a factor in innovation. Dougherty (1988) shows the power of a cultural analysis in identifying the origins and effects of poor interfunctional cooperation in new product development. Section 2.1.3 suggested the importance of cultural issues in defining mission and Section 2.3.2 will comment on their importance in sharing across business units.

Third, there is a continuing stream of literature on the management of inter-functional relations (see bibliography 2.1.4.9). Here, apart from the work of Dougherty cited in the previous paragraph, there is the considerable body of research by Rubenstein and his colleagues at Northwestern University's Program on the Management of Research, Development and Innovation (see Rubenstein and Ginn, 1985). The organizational interfaces are among the more complex issues of technology management; but the quality of interface management clearly conditions both the implementation of technology strategy and the information flows and climate for innovation.

Finally, the literature on ''intrapreneurship'' has generated important new insights into the relationship between organization design and technology strategy. Efforts to bring the creativity and responsiveness of small firms into large corporations have offered a propitious terrain for technology strategy research. The results contribute to the ''new lines of business'' policy material (bibliography 2.1.4.10b).

These bodies of research on structures for innovation reorient us towards strategy formulation and thus partially close the strategy formulation-implementation-reformulation loop. Given the uncertainty surrounding the robustness of any technology strategy, the organizational structure that implements this strategy plays a critical role in assuring the ability of the organization to refine or redefine its strategic direction. Structure is both the result of strategy choices (Chandler, 1962) and the context conditioning future strategy choices (Burgelman, 1986). In technology strategy just as in business strategy, implementation cannot be relegated to a secondary role.

2.3 The Technology Strategy Process

As already indicated, for researchers of the "incrementalist" school such as Quinn (1986, 1988), the strategy process is its own content. Even for those who would not go this far, it is clear that many of the steps in strategy formulation—like STA and DTC analysis—serve an important cognitive purpose even before they are put to their ultimate analytic use. In this section, I shall very briefly review the technology strategy process in two contexts, first technical entrepreneurship and second, the more formal process characteristic of larger firms.

In exploring these issues, the literature on the business strategy process is the natural starting-point. But a focus on technology strategy reveals that it is difficult to isolate the process issues from the content issues. Clarity on the nature of the technology strategy process calls for greater interaction between research on strategy process and content (Huff and Reger, 1987).

2.3.1 Technical Entrepreneurship

Entrepreneurial "high-tech" firms, by virtue of both the centrality of technology in determining their competitiveness and the importance of technological change in continually reshaping their environment, are potentially rich in lessons for technology strategy. They provide a critical reference point for the strategy process literature because of their congeniality to creativity and their responsiveness to the market, in contrast with the all-too-common bureaucracy and inertia of larger firms. As pointed out by Bahrami and Evans (1988), however, most of the strategy process literature has been based on public sector organizations (Lindlom, 1959; Allison, 1971), educational institutions (March and Olson, 1976), nonprofit organizations (Mintzberg and McHugh, 1985), or large diversified firms in relatively stable environments (Chandler, 1962; Bower, 1970; Quinn, 1985; Mintzberg and Waters, 1982). Research on the technology strategy process in high-technology firms is just beginning to accumulate (Bourgeois and Eisenhardt, 1987, 1988; Bahrami and Evans, 1987, 1988; Romanelli, 1987). The key characteristics of the technology strategy process in these firms appear to be related to the need for speed and flexibility. In volatile technology-intensive contexts, action is often a prerequisite for strategy formulation.

But research on the strategy process in these firms is difficult: the nature of the process is difficult to detach from small size, newness, and the entrepreneurial personality of the founders. The research in this field suffers from the fact that many of the key mechanisms are thus invisible to the researcher, since they are buried in the minds of the actors, rather than objectified in the explicit procedures often found in larger organizations. Moreover, the weight of entrepreneurial ideology is heavy in this literature. This section's bibliography avoids purely celebratory literature, but little has reached the sophistication called for by Penrose (1980).

2.3.2 The Technology Strategy Process in Larger Firms

If we turn now to the literature on the strategy process in larger firms, one of the key lessons appears to be that the cycle linking formulation, implementation and evaluation is of necessity a political process (Pettigrew, 1973; Pfeffer and Salancik, 1974; Mintzberg, 1983). Even if many of the actors are convinced that their own interests are best served by the prosperity of the organization, the process will be characterized by different views on the best way of achieving that prosperity—and the downward slope to factionalism is a slippery one. In the case of small entrepreneurial companies, this political process is masked, and appears in the form of ambivalences in the thinking of the entrepreneur or in the form of conflictual interpersonal relations within a small management team. In larger organizations with greater role differentiation the effectiveness of the technology strategy process depends critically on the political conditions enabling or impeding dialogue across functions and across management levels (Bower, 1970; Doz et al., 1986).

To say that the strategy process is intensely political is not, however, to say that it is only political. The properly cognitive dimensions have already been mentioned, and there is clearly an important and intimate relationship between the strategy process and the culture, or constellation of values, characteristic of the organization. Moreover, the strategy process will reflect both external and internal economic pressures. Objective external resource constraints shape the strategy process, as does the players' ability to understand those constraints. Internal economic pressures—the different incentives experienced by different business-unit and functional managers—will also color the strategy dialogue. Finally technology trends and opportunities will shape not only strategy content but also the strategy agenda and the roles played by different players in the strategy process. This complexity has been nicely demonstrated by research such as Burgleman's on corporate intrapreneurship (see bibliography 2.1.4.10b).

Within this complex structure of interacting forces, a broad spectrum of research on the technology process in larger firms suggests that the political and symbolic forces are particularly potent. (For a more theoretical argument concerning the relative efficacy of technological, economic, political and symbolic forces on strategy, see Adler and Borys, 1988). Nowhere is the political and symbolic character of the strategy process more apparent than where divisions in a large multidivisional firm need to share technological resources. The associated policy issues are discussed in literature under bibliography 2.1.4.10a. Here I would like to highlight the process issues and their origins.

Eccles (1985) has argued that when the business units of a multidivisional company are somewhat interdependent—that is, they are neither totally dependent on each other as in the vertical integration case nor totally independent as in the unrelated diversification case—corporations typically have great trouble identifying a transfer pricing system that will give the right incentives to division general managers. The discussion in Section 1.1 of the dual nature of technology helps us understand the source of this difficulty.

Extending into the multidivisional firm the comments made in Section 1.1 concerning technology's explicit component, it can be shown that when divisions are interdependent because they rely on a common resource that has some of the characteristics of a "public good," then there can be no optimal transfer price (Arrow and Hurwicz, 1977). If transfers are based on market price, they will tend to optimize the incentive to produce this common resource; but from the corporation's viewpoint these prices will unduly restrict the resource's availability to other business units. If transfer is based on cost, then distribution will be optimal, but there will be little incentive for the division that is otherwise measured on profitability to take the risk necessary to develop new technology. Extending into the multidivisional firm the discussion in Section 1.1 of technology's tacit dimension reinforces the conclusion: neither transfer prices nor corporate mandate can ensure an optimal pattern of technology development and diffusion within the corporation.

Ouchi's (1980) argument can be read as an extension of this analysis: neither centralized hierarchical control, nor divisional autonomy, nor even some astute combination of the two, can match the power of combining appropriate administrative and economic mechanisms with a substantial organizational investment in shared non-pecuniary values—building shared commitments to the organization's superordinate goals and thus enabling both aggressive development and open sharing of new technologies. In Ouchi's terminology, when goals and performance are ambiguous—as is typically the case in technology development projects—then "clans" will outperform both hierarchical and market forms of governance.

The experience of some larger multidivisional companies seems to confirm the need for a full spectrum of team-building mechanisms—economic, administrative and cultural—for creating and sustaining what Porter (1985) calls "horizontal strategy" (for a description of the TRW case, see *Business Week,* 1983). A study by Horwitch and Thiertart (1987) begins the task of identifying the types of business-unit interdependencies most appropriate in various contexts.

2.4 Technology Opportunity and Strategy

The preceding subsections have identified a range of organization-level factors at play in the determination of technology strategy. In concluding this section, however, it is worth recalling the fundamental uncertainty surrounding this determination introduced by technological change itself.

Business strategy is a complex, open-ended game, where changing market preferences and competitor moves undermine the best laid plans. Technology adds a distinctive layer of uncertainty to this already uncertain process, through its unpredictably dynamic evolution (see Section 1.1).

But there is a second type of uncertainty associated with technology, an uncertainty due to our current lack of conceptual understanding of even the more enduring features of technology. Section 2.1.3. referred to the first step in technology strategy determination as SWOT—the identification of the organization's technological strengths, weakness, opportunities and threats. Section 1 concluded that we needed more robust ways of characterizing the structure of the technological opportunity space. Section 2 reinforced this point: at the organization level, strategy formulation needs not only a robust characterization of the environment's technology opportunities but also a substantive characterization of the organization's technological focii. This may be an area where progress in research will depend on progress in industry practice.

3. PROJECTS

3.1 Introduction

There are several possible ways to approach the study of innovation projects. The previous section's discussion of the technology policy domains has drawn extensively on these literatures:

- Bibliography 3.1.1 lists several valuable guides to the literatures on the technology innovation process as well as general studies in this area. I have included in this section references on the important subject of key roles in the innovation process.
- Bibliography 3.1.2 focuses on the determinants of innovation success and failure. In this literature, one finds valuable discussion of the importance of the marketing interface and many of the project management issues discussed in Section 2.1.4.7.
- Bibliography 3.1.3 focuses on the information flows underlying the innovation process—this literature encompasses flows both internal to the company (relevant to Sections 2.1.4.2,9,10) and external (Section 2.1.4.5).
- Bibliography 3.1.4 lists some of the key references that attempt to generalize across innovations to characterize the diffusion process—it covers the fields discussed under the policy sections on sources and commercialization of technology (2.1.4.5).
- Bibliography 3.1.5 lists some entry points into the literature on technology transfer—it encompasses both the inter-organizational and inter-divisional issues discussed under Sections 2.1.4.5,10.

It is striking however, that neither the extensive National Science Foundation review of the innovation literature (Tornatsky et al., 1983) nor the Sussex University Science Policy Research Unit's research bibliography (Henwood and Thomas, 1984) have a chapter on strategy. As I argued in the introduction, this reflects the dominance of the single innovation project as the unit of analysis in this literature.

I have taken two complementary approaches to identifying the potential strategy lessons of the literatures on innovation projects. The previous section drew on the innovation literatures to help characterize technology management's key strategy and policy issues. This third section suggests two ways in which strategy is embedded in individual technical projects. This will lead us to identify some key shortcomings in the research to date.

3.2 Strategic and Operating Issues in Managing Technical Projects

There are two key ways in which strategy is embedded in projects. First there are two phases of project management whose raison d'être is linkage to strategy—a pre-project phase and a post-project phase. Second, all the other project phases can also be shown to implicate strategy. I shall address these two issues in turn.

First, it is unfortunate that most phase models ignore the pre- and post-project phases. Kelly and Kranzberg's innovation process model, for example, distinguishes (1) problem definition and idea generation, (2) invention of prototype, (3) research and development, (4) application, and (5) diffusion into new contexts. (See Kelly and Kranzberg, 1978, Tornatzky et al., 1983, ch. II, and Pelz, 1983, for discussions of stage models of innovation). Explicit identification of these two extra phases can, however, help bridge the strategic level and the project level of analys:

- The *pre-project phase* encompasses the activities by which the firm establishes its priorities and identifies the technologies it expects to be involved in future projects. Useful strategic insights for the management of this activity might include ideas like mapping product and process technology evolution and separating as much as possible the development of technologies from the development of products or processes using these technologies so as to minimize the risk of stalling the project (Hayes, Wheelwright, and Clark, 1988, ch. 10). Amongst the other useful starting points for research here are: Cooper's (1984a,b) exploration of new product strategies; Gold's (1981) discussion of the "pre-decision environment" of innovation adoption; Bower's (1970) discussion of the "structural context" set by senior management for inhibiting, stimulating and challenging bottom-up project formulation; and Burgelman's (1983) discussion of the "strategic context" linking specific projects to explicit and emergent strategy. (For these and other relevant references, see bibliography 3.2.1).
- The *post-project phase* governs the ability of the organization to learn from one project to the next, to assess the effectiveness of the organizational or technical approaches employed, and to make improvements that, over a series of projects, will leverage project experience into a more innovative system (Maidique and Zirger, 1985; Imai, 1986). One element of such a learning system might be an engineering change tracking/evaluation system; another might be planning for staffing critical project management positions to balance the need to accumulate experience with the need to diffuse this scarce expertise. The literature on organizational learning is broad (see the reviews by Hedberg, 1981; Fiol and Lyles, 1985;

> Levitt and March, 1988); it could usefully be focused on this process. Weick's (1979) focus on retrospective rationalization is also useful in understanding how organizations interpret and learn (or don't learn) from project experience. (For these and other relevant references, see bibliography 3.2.2).

As these characterizations make clear, the point is not to hypothesize some artificially linear innovation process. Clearly, within a given project there is often iteration and advance scanning of later phases. Moreover, in organizations with more than one project underway, at any given point in time different parts of the organization may be at work on different phases of different projects. But the strategic leverage of recognizing pre- and post-project activities as logically distinct phases can be considerable.

The second way strategy is embedded in projects is more pervasive: a moment's reflection suffices to show that every phase consists of activities involving both strategic and tactical issues. Strategic issues in project management differ from operating or tactical issues in that the former are longer-term in their time horizon, planning-oriented rather than execution-oriented, and designed to reduce critical uncertainties rather than managing the remaining uncertainties. Many of the policy domains listed in Exhibit 5—and not only those listed under managing technical projects—are both reflected in and modified by project activity.

Take as an example the problem-definition phase. This phase involves activities such as ensuring a focus on high-leverage areas and coupling downstream, and these activities involve both strategic and tactical issues. Ensuring a focus on high-leverage areas involves strategic issues such as interfunctional relations with marketing and external relations with information sources, as well as tactical issues such as the use of brainstorming techniques, nurturing gatekeepers and avoiding the "not invented here" syndrome. Downstream coupling involves strategic issues such as project team composition and reporting relationships, as well as tactical ones such as how to balance subunits' priorities and influence, how to fill critical sponsor roles, and the timing and depth of reports and reviews. (Another, more theoretical way to think of the relationship between what I am calling strategic and tactical issues might be the "structurationist" framework proposed by Giddens, 1979, in which "structure"—the organized rules and resources of the organization—constitutes the conditions of "action"—the knowledgeable activity of social actors—and action is the medium through which structure is both reproduced and transformed.)

The lesson of research into the nature of managerial work (Mintzberg, 1973, 1975) is clear: strategic and operating issues are both present in most activities undertaken in the manager's workday. Moreover, it is through and by their involvement in the operating issues that most managers both refine their strategic maps and influence the strategic orientation of their organization (Kotter, 1988; Isenberg, 1987). These observations undermine any simplistic strategy-versus-operations distinction; but they also underline the value of research designed to help us understand the complex interrelationship between strategic and operating issues.

The linkage between strategy and projects is of vital significance: projects are the form in which new technologies are developed and implemented. If technology strategy is, at its heart, a capabilities development vector, then it is through projects that it operates. Implementation projects draw on and refine, and development projects contribute to, the knowledge-base of the organization (Jaikumar, 1986); technology strategy is a map of the desired evolution of that knowledge-base. "Good project management" thus consists in part of the capacity to align specific project activities with the project mission, and these two with the organization's goals and its strengths, weaknesses, opportunities and threats (extending Fujimura, 1987).

3.3 Strategic Lessons

An examination of the literatures on technological innovations highlights several difficulties in drawing strategic lessons from the innovation process research:

- The project management literature has suffered from a preoccupation with the tactical aspects. This preoccupation is due to two factors. First, these aspects are easier to formalize than the strategic issues. Second, the project and its organizational context are often treated as two distinct entities that only communicate via an initial definition by the organization of the objective and a final delivery of the product by the project team. Dickson (1986) highlights the need for closer organization/project relations both for better project management practice and for a more realistic research focus.
- Much of the more analytic work on R&D management has been preoccupied by the construction of algorithms for project selection, and little has adopted a system-analytic approach which might be more powerful in capturing other strategic issues (Roberts, 1962).

- The new product success and failure literature is too focused on individual products and thus generates little by way of insight into strategic, as distinct from tactical, issues (Radnor and Rich, 1980; Maidique and Zirger, 1985).
- The diffusion literature is too often focused on individual adopters, as opposed to organizational adopters: both tactical and strategic issues are probably significantly different across these different contexts (Kelly and Kranzberg, 1978, pp. 119ff; Rogers and Eveland, 1978).

Apart from these specific difficulties, I would suggest another, more fundamental and common difficulty—one that has already surfaced in the two preceding sections of this essay: the research on the innovation process has not yet developed the theory that would allow it to distinguish technology management challenges according to the type of technology being developed or deployed. Most of the research in each of the literatures I have referred to in this section has sought to identify lessons for innovation management that would apply equally well to all technology contexts. But numerous studies have made it clear that this level of generalization is not likely to be very useful. (Downs and Mohr, 1976, and Kimberly and Evanisko, 1981 raise this concern with respect to the diffusion literature; Goodman, 1986, raises it with respect to the work group effectiveness literature).

Many researchers have attempted to deal with this challenge by using generic typologies of tasks. Some have thus referred to the importance of the degree of "technical complexity" in project management (J. R. White, 1982). Others have discussed how greater "uncertainty" in the evaluation of costs or benefits necessitates a less formal approach to evaluation of automation projects (Dean, 1987). Yet others discuss the difference between the organizational requirements of "incremental" and "radical" innovations in manufacturing (Ettlie, Bridges and O'Keefe, 1984). (See Goodman 1986 for a review of task and technology typologies).

But these very simple and abstract characterizations do not seem to capture enough of the specificity of tasks and technologies required to make sense of technology management practice. Other researchers have therefore pushed toward greater refinement:

- Abernathy and Utterback (1982) identify distinct technology lifecycles for product and process technologies and identify distinct management issues for each phase. Moore and Tushman (1982) discuss comparable differences between stages in the product life-cycle.

- The management of research teams has been contrasted with that of development teams (Abetti, 1985; Allen, Lee, Tushman, 1980; Allen, Tushman, Lee, 1979; Blandin and Brown, 1977; Cleland and King, 1975; Cooper, 1983; Katz and Tushman, 1979, 1981; Keller and Holland, 1975; Leifer and Triscari, 1987; Tushman, 1977, 1978, 1979b; Twiss 1980; W. White, 1982).
- Rosenbloom and Wolek (1970) distinguish project information sources according to whether the project is expected to make small or large contributions to the advance of knowledge (design versus research or development) and whether the project is expected to make small or large contributions to the conduct of operations (research versus design or development).
- Some important work was conducted in the 1960s on the micro-economics of R&D, in which it was shown that component state-of-the-art advance and component interrelatedness should influence the choice between parallel and sequential development policies (Marchak et al., 1967, ch 2.)

But even these typological distinctions do not seem to get far enough into the differences between technology domains. Leifer and Triscari (1987) note that much of the above-mentioned research by Allen and Tushman was based on their study of a chemical products company, and that some of the differences between research and development teams in this company did not obtain in Leifer and Triscari's study of several military technology development teams. Similarly, one cannot help wondering if there are any systematic differences in the nature of technology development tasks across disciplines seeing the greater frequency of multiple-author papers in computer hardware than software (Subramanyam, 1983; see also Small and Griffin, 1974) or seeing the quite different managerial precepts advanced in the literature on hardware and software development (Hauptman, 1988). But further, within technical disciplines, and indeed within their research phase, what are the differences between projects of an "obsoleting" kind (moving back up the design hierarchy) and those of a "refining" kind (moving "down" the hierarchy)? In the new field of design theory, research has focused on the difference between expert and novice designers within a given technical field, but differences across fields have not yet been sufficiently researched (Rouse and Boff, 1987). The unfortunate conclusion appears to be inescapable: we know too little about the dimensions of the technical task and how these dimensions influence project management.

The lack of concepts with which to characterize the structure of the technical task might help explain such phenomena as the disappointingly slow diffusion of project management techniques. Defining the activities that need to be sequenced is the major hurdle that must be surmounted before the value of better techniques for scheduling those activities (PERT, etc.) can be appreciated; but we lack the concepts that can help project managers identify the key activities. Project managers in different fields accumulate tacit knowledge on the challenges of their task, but we have no theoretical framework to make that knowledge explicit and to thus make it amenable to systematic improvement.

In part, this gap can be explained by the natural desire of management theorists to generalize across as broad a variety of technologies as possible. In some sub-areas of research, such generalizations may generate important and useful insights. In the vast bulk of the research on technological innovation, however, it is difficult to avoid the impression that generalizations across technology types miss many of the key issues.

We thus conclude this section with an extension of the conclusions of sections 1 and 2: technology opportunities, technology direction and technical tasks are poorly understood. One potentially rich avenue for exploring technical tasks and the associated project-level analysis of technology dynamics is provided by cognitive science: thinking of technology not as equipment but as the knowledge underlying that equipment might allow researchers to deploy a rich repertoire of concepts from cognitive development theories (see for example, Jaikumar and Bohn, 1986; Levitt and March 1988; Clark and Fujimoto, 1987; Daft and Lengel, 1986). The juxtaposition of cognitive science concepts and grounded-theory methodology seems to offer a promising avenue for research in this crucial area.

4. CONCLUSION

This review of the three levels—environment, organization and project—has identified numerous strategic issues surrounding technology. The accelerating pace of technological change suggests that the strategic management of technology will, over time, become more important for a broader spectrum of organizations.

The effort to manage accelerating technological change strategically creates opportunities and pressures that seem to spell a profound transformation of the firm:

- As the pace of change accelerates, a greater proportion of managerial mind-share will need to be devoted to strategic, as distinct from operating, issues.
- Simultaneously, the increasing pace of change will tend to create an information-processing overload at the higher levels of management unless some strategy work is delegated to lower-level participants such as functional and project managers.
- And finally, senior management, notwithstanding this delegation, will need to become more technologically sophisticated in order to lead the resulting technology strategy dialogue.

In an increasingly technologically dynamic world, the potential for improved performance through better technology strategy focus at all levels in the organization suggests that a multifunction, multilevel concept of technology strategy will become increasingly central to competitiveness. Such is the message of a recent National Research Council report on ''Management of Technology: The Hidden Competitive Advantage'' (1987).

The need for a concept of technology strategy is thus destined to intensify. How far have researchers come in meeting it? In the analysis of the organizational level itself, there has been an impressive accumulation of normative, pragmatic models of technology strategy and some important results from more narrowly focused theoretical research; but the bridge between the two types of literature is weak. While research on both the environmental context and innovation projects is extensive, a technology strategy framework calls for some reconceptualizations (especially in the environment domain) and some extensions (especially in the innovation projects domain). In all three levels of the framework, my discussion has highlighted the potential value of greater insight into the managerially-significant dimensions of technologies themselves.

Nine years ago, Kantrow, reviewing the literature on the strategy-technology connection (1980) wrote: ''The major unfinished business of the research literature is to provide managers with needed guidance in their formulation of a technological strategy for their companies.'' His overall assessment of the field was that ''the most basic categories and terminology of technology strategy have not yet been satisfactorily determined.'' This chapter suggests that Kantrow's assessment is still valid.

The general thrust of this essay has been that the study of technology strategy is not a discipline with a well-developed internal motor of conceptual development. On the contrary, research in this field necessarily trails behind practice. At its best, research in the form of in-depth case studies and rigorous historical analysis allows us to confront industry practice with theoretical models drawn from disciplines such as economics and sociology in order to creatively synthesize the lessons of industry.

ACKNOWLEDGMENT

This article has benefited from the comments of Robert Burgelman, Bob Hayes, Mel Horwitch, Bob Keeley, Bill MacDonald, Tom McAvoy, Richard Rosenbloom and Steven Wheelwright.

BIBLIOGRAPHY

0. INTRODUCTION

Andrews, K. R., *The Concept of Corporate Strategy,* Homewood, Il: Richard D. Irwin, 1980.

Cohen, W. M. and Mowery, D. C., "Firm Heterogeneity and R&D: An Agenda for Research," in B. Bozeman, M. Crow, A. Link, eds., *Strategic Management of Industrial R&D,* Lexington, MA: Heath, 1984.

Friar, J. and Horwitch, M., "The Emergence of Technology Strategy: A New Dimension of Strategic Management," in M. Horwitch, ed., *Technology in Society,* 7, 2/3, 1986, pp. 50–85.

Gold, B. "Rediscovering the Technological Foundations of Industrial Competitiveness," *Omega,* 8, 5, 1980, pp. 503–504.

Kelly, P. and Kranzberg, M., eds., *Technological Innovation: A Critical Review of Current Knowledge,* San Francisco: San Francisco Press, 1978.

Maidique, M. A. and Zirger, B. J., "New Products Learning Cycle," *Research Policy,* December 1985, pp. 1–40.

Mintzberg, H., "The Strategy Concept I: Another Look at Why Organizations Need Strategy," *California Management Review,* Fall 1987, pp. 25–32.

Mintzberg, H., "The Strategy Concept I: Five Ps for Strategy," *California Management Review,* Fall 1987, pp. 11–24.

Radnor, M. and Rich, R. F., "Organizational Aspects of R&D Management: A Goal-Directed Contextual Perspective," *TIMS Studies in the Management Sciences,* 15, 1980, pp. 113–133.

Rosenbloom, R. S., "Technological Innovation in Firms and Industries: An Assessment of the State of the Art," in P. Kelly and M. Kranzberg, eds., *Technological Innovation: A Critical Review of Current Knowledge,* San Francisco: San Francisco Press, 1978.

Scott, J. T., "Firm Versus Industry Variability in R&D Intensity," in Z. Griliches, ed., *R&D, Patents, and Productivity,* Cambridge, MA: National Bureau of Economic Research, 1984.

Utterback, J. M., "Innovation and Corporate Strategy," *International Journal of Technology Management,* 1, 1/2, 1986.

Wheeler, T. L. and Hunger, J. D., *Strategic Management and Business Policy,* Reading, MA: Addison-Wesley, 1986.

1. THE ENVIRONMENT

1.1 Technology dynamics

Abernathy, W. J. and Clark, K. B., "Innovation: Mapping the Winds of Creative Destruction," *Research Policy,* 14, 1985, pp. 3–22.

Abernathy, W. J. and Townsend, P., "Technology, Productivity and Process Change," *Technological Forecasting and Social Change,* 7, 19, 1975, pp. 379–396.

Abernathy, W. J., Clark, K. B. and Kantrow, A. M., *Industrial Renaissance,* New York: Basic Books, 1983.

Abernathy, W. J., *The Productivity Dilemma: Roadblock to Innovation in the Automobile Industry,* Baltimore: Johns Hopkins University Press, 1978.

Alexander, C., *Notes on the Synthesis of Form,* Cambridge MA: Harvard University Press, 1964.

Anderson, C. R. and Zeithaml, C. P., "Stage of the Product Life Cycle, Business Strategy, and Business Performance," *Academy of Management Journal,* 27, 1, 1984, pp. 5–24.

Ansoff, H. I., *Implanting Strategic Management,* Englewood Cliffs, NJ: Prentice-Hall International, 1984.

Arrow, K. J., "Economic Welfare and the Allocation of Resources for Invention," in *The Rate and Direction of Inventive Activity,* Princeton: Princeton University Press for the National Bureau of Economic Research, 1962.

Arthur, W. B., "*Competing Techniques and Lock-in by Historical Events. The Dynamics of Allocation Under Increasing Returns,*" CEPR, Stanford University, 1985.

Baily, M. N. and Chakrabarti, A. K., "Innovation and Productivity in U.S. Industry," *Brookings Papers on Economic Activity,* 2, 1985, pp. 609–639.

Bhide, A., "Hustle as Strategy," *Harvard Business Review,* Sept–Oct. 1986, 5, pp. 59–65.

Birnbaum, P. H., "Strategic Management of Industrial Technology: A Review of the Issues," *IEEE Transactions on Engineering Management,* EM-31, 4, Nov 1984, pp. 186–191.

Braunstein, Y. M., Baumol, W. J. and Mansfield, E., "The Economics of Innovation," *TIMS Studies in the Management Sciences,* 15, 1980, pp. 19–32.

Clark, K. B. and Griliches, Z., "Productivity Growth and R&D at the Business Level: Results from the PIMS Data Base," in Z. Griliches, ed., *R&D, Patents and Productivity,* Chicago, IL: University of Chicago Press, 1984.

Clark, K. B., "Competition, Technical Diversity and Radical Innovation in the U.S. Auto Industry," in R. S. Rosenbloom, ed., *Research on Technological Innovation, Management and Policy,* 1, 1983, pp. 103–149.

Clark, K. B., "The Interaction of Design Hierarchies and Market Concepts in Technological Evolution," *Research Policy,* 14, 1985, pp. 235–251.

Cooper, A. C. and Schendel, D. "Strategic Responses to Technological Threats," *Business Horizons,* Feb. 1976, pp. 61–69.

David, P. A., *Technical Choice, Innovation and Economic Growth,* Cambridge UK: Cambridge University Press, 1975.

David, P. A., "Clio and the Economics of QWERTY," *American Economics Review,* 75, 2, May 1985, pp. 332–337.

De Bresson, C. and Lampel, J., "Beyond the Life Cycle Part II. An Illustration," *Journal of Product Innovation Management,* 3, 1985, pp. 188–195.

De Bresson, C. and Lampel, J., "Beyond the Life Cycle: Organizational and Technological Design Part I. An Alternative Perspective," *Journal of Product Innovation Management,* 3, 1985, pp. 170–187.

Dosi, G., "Sources, Procedures and Microeconomic Effects of Innovation," *Journal of Economic Literature,* Sept 1988, pp. 1120–1171.

Dosi, G., *Technical Change and Industrial Transformation,* London: MacMillan, 1984.

Dosi, G., "Technological Paradigms and Technological Trajectories: A Suggested Interpretation of the Determinants and Directions of Technical Change," *Research Policy,* June 1982, pp. 147–162.

Durand, T. and Gonard, T. "Management de la Technologie: Stratégie des Firmes face à une Rupture Technologique dans l'Industrie de l'Insuline," *Revue Française de Gestion,* Dec. 1986.

Forrester, J. W., "Innovation and the Economic Long Wave," *Management Review,* June 1979.

Foster, R. N., "Boosting the Payoff from R&D," *Research Management,* January 1982.

Foster, R. N., *Innovation: The Attacker's Advantage,* New York: Summit, 1986.

Freeman, C., Clark, J. and Soete, L., *Unemployment and Technical Innovation: A Study of Long Waves and Economic Development,* London: Frances Pinter, 1982.

Freeman, C., ed., *Long Waves in the World Economy,* Sevenoaks: Butterworths, 1983.

Freeman, C., *The Economics of Industrial Innovation,* 2nd ed., Cambridge MA: MIT Press, 1986.

Gershuny, J., *Social Innovation and the Division of Labour,* Oxford: Oxford University Press, 1983.

Gibbons, M. and Johnson, R. D., "The Role of Science in Technological Innovation," *Research Policy,* Vol. 3, No. 3, Nov 1974, pp. 220–242.

Gold, B., ed., *Research, Technological Change and Economic Analysis,* Lexington, MA: D. C. Health, 1977.

Gold, B., ed., *Technological Change: Economics, Management and Environment,* Oxford: Pergamon Press, 1975.

Habakkuk, H. J., *American and British Technology in the Nineteenth Century,* Cambridge: Cambridge University Press, 1962.

Hirsh, R. F., "How Success Short Circuits the Future," *Harvard Business Review,* March–April, 1986.

Hunter, M. W., II., "Are Technological Upheavals Inevitable?" *Harvard Business Review,* Sept–Oct. 1969.

Ketterington, J. and White, J., "Making Technology Work for Business," in R. Lamb, ed., *Competitive Strategic Management,* Englewood Cliffs NJ: Prentice-Hall, 1984.

Kodama, F., "Japanese Innovation in Mechatronics Technology," *Science and Public Policy,* 13, 1, Feb 1986a.

Kodama, F., "Technological Diversification of Japanese Industry," *Science*, 233, July 18, 1986b.

Kondratief, N. D., "Long Waves in Eonomic Life," *Lloyds Bank Review*, July 1978.

Kuhn, T. S., *The Structure of Scientific Revolutions* 2nd ed., Chicago: University of Chicago Press, 1970.

Kusterer, K., *Know How on the Job*, Boulder, CO: Westview, 1978.

Landes, D., *The Unbound Prometheus*, Cambridge UK: Cambridge University Press, 1969.

Levin, R. C., Cohen, W. M. and Mowery, D. C., "R&D Appropriability, Opportunity and Market Structure: New Evidence on Some Schumpeterian Hypotheses," *American Economic Review*, May 1985, pp. 20–24.

Levin, R. C., Klevorick, A. K., Nelson, R. R. and Winter, S. G., "Appropriating the Returns from Industrial Research and Development," *Brookings Papers on Economic Activity*, 3, 1987, pp. 783–831.

Malerba, F., *The Semiconductor Business: The Economics of Rapid Growth and Decline*, Madison: University of Wisconsin Press, 1985.

Mandel, E., "Explaining Long Waves of Capitalist Development," *Futures*, 4, August 1981, pp. 332–338.

Manwaring, T. and Wood, S., "The Ghost in the Labour Process" in D. Knights, H. Willmott, D. Collinson, eds., *Job Redesign: Critical Perspectives on the Labour Process*, London: Gower, 1965.

Mensch, G. O., *Stalemate in Technology*, Cambridge MA: Ballinger 1979.

Mowery, D. C., "The Relationship Between Intrafirm and Contractual Forms of Industrial Research in American Manufacturing, 1900–1940," *Explorations in Economic History*, 20, 4, Oct 1983, pp. 351–374.

Nelson, R. R. and Winter, S. G., *An Evolutionary Theory of Economic Change*, Cambridge, MA: Harvard University Press, 1982.

Nelson, R. R. and Winter, S. G., "In Search of a Useful Theory of Innovations," *Research Policy*, Jan 1977.

Pavitt, K., "*Chips and Trajectories: How Will the Semiconductor Influence the Sources and Directions of Technical Change?*" Paper presented at Stanford TIP Workshop, March 1985.

Pavitt, K., Robson, M. and Townsend, J., "*Technological Accumulation, Diversification and Organization in UK Companies, 1945–1983*," Science Policy Research Unit, University of Sussex, Discussion Paper no. 50, 1987.

Pavitt, K., Robson, M. and Townsend, J., "The Size Distribution of Innovative Firms in the UK: 1945–1983," *Journal of Industrial Economics*, 3, 35, March 1987, pp. 297–319.

Pavitt, K., "Sectoral Patterns of Technical Change: Towards A Taxonomy and A Theory," *Research Policy*, 13, 1984.

Polanyi, M., *The Tacit Dimension*, Garden City, NY: Doubleday Anchor, 1967.

Price, D. J. de S., "The Science/Technology Relationship, the Craft of Experimantal Science, and Policy for the Improvement of High Technology Innovation," *Research Policy*, 12, 1, Jan 1984.

Price, D. J. de S., *Little Science, Big Science and Beyond*, New York: Columbia University Press, 1986.

Rink, D. R. and Swan, J. E., "Product Life Cycle Research: A Literature Review," *Journal of Business Research,* Sept. 1979, pp. 219–242.

Rosenberg, N. and Frischtak, C. R., "Technological Innovation and Long Waves," *Cambridge Journal of Economics,* 8, 1984, pp. 7–24.

Rosenberg, N., *Inside the Black Box: Technology and Economics,* Cambridge, MA: Cambridge University Press, 1982.

Rosenberg, N., *Perspectives on Technology,* Cambridge, MA: Cambridge University Press, 1976.

Rosenbloom, R. S. and Freeze, K. J., "Ampex Corporation and Video Innovation," in R. S. Rosenbloom, ed., *Research on Technological Innovation Management and Policy,* 2, 1985.

Roussel, P. "Technological Maturity Proves A Valid and Important Concept," *Research Management,* Jan–Feb. 1984, pp. 29–34.

Sahal, D., "Alternative Conceptions of Technology," *Research Policy,* Vol. 10, 1981a, pp. 2–24.

Sahal, D., *Patterns of Technological Innovation,* New York: Addison-Wesley, 1981b.

Sahal, D., "Technological Guideposts and Innovation Avenues," *Research Policy,* 14, 2, April 1985, pp. 61–82.

Schmookler, J., *Invention and Economic Growth,* Cambridge, MA: Harvard University Press, 1966.

Schumpeter, J. A., *The Theory of Economic Development,* Cambridge, MA: Harvard University Press, 1949.

Stobaugh, R. B., *Innovation and Competition,* Boston, MA: Harvard Business School Press, 1988.

Teece, D. J., "Profiting from Technological Innovation," *Research Policy,* 1986, pp. 285–306.

Tushman, M. L. and Anderson, P., "Technological Discontinuities and Organizational Environments" *Administrative Science Quarterly,* 31, 1986, pp. 439–465.

Usher, A. P., *A History of Mechanical Inventions,* Harvard University Press, 1954.

Utterback, J. M. and Abernathy, W. J., "A Dynamic Model of Product and Process Innovation," *Omega,* 3, 6, 1975.

Van Duijin, J. J., "Fluctuations in Innovation Over Time," *Futures,* 4, August 1981, pp. 246–275.

Van Duijin, J. J., "The Long Wave in Economic Life," *De Economist,* 4, 1977, pp. 543–576.

Von Hippel, E., "The Dominant Role of the User in Semiconductor and Electronic Subassembly Process Innovation," *IEEE Transactions on Engineering Management,* EM-24, 2, May 1977, pp. 60–71.

Wilkinson, A., "Technology—An Increasingly Dominant Factor in Corporate Strategy," *R&D Management,* 13, 4, 1983, pp. 245–259.

1.1.1 Technological Forecasting

Ayres, R. U., *Technological Forecasting and Long Range Planning,* New York: McGraw-Hill, 1969.

Bright, J. R., "Evaluating Signals of Technological Change," *Harvard Business Review*, Jan–Feb. 1970.

Bright, J. R., *Practical Technological Forecasting*, Austin, TX: Industrial Management Center, 1978.

Bright, J. R., "Technological Forecasting Literature: Emergence and Impact on Technological Innovation," in P. Kelly and M. Kranzberg, eds., *Technological Innovation: A Critical Review of Current Knowledge*, San Francisco: San Francisco Press, 1978.

Cetron, M. J., *Technological Forecasting*, New York: Gordon and Breach, 1969.

Fusfeld, A. R. and Spital, F. C., "Technology Forecasting and Planning in the Corporate Environment: Survey and Comment," in B. V. Dean and J. L. Goldhar, eds., *Management of Research and Innovation, TIMS Studies in the Management Sciences*, 15, 1980, pp. 151–162.

Harris, N. D., "Forecasting Military Requirements: A Critical Viewpoint of the Industry's Approach," *IEEE Transactions on Engineering Management*, EM-19, 1, 33, Feb 1972.

Jantsch, E., *Technological Forecasting in Perspective*, Paris: OECD, 1967.

Jones, H. and Twiss, B. C., *Forecasting Technology for Planning Decisions*, London: Macmillan, 1978.

Makridakis, S., Wheelwright, S. C., and McGee, V. E., *Forecasting: Methods and Applications*, New York: John Wiley and Sons, 1983.

Mansfield, E., *Industrial Research and Technological Innovation*, New York: W. W. Norton, 1968.

Martino, J. P., *Technological Forecasting for Decision Making*, New York: North-Holland, 1983.

Nutt, A. B., "Technological Forecasting in an R&D Laboratory," *IEEE Transactions on Engineering Management*, EM-19, 1, 34, Feb 1972.

Porter, A. L., Rossini, F. A., Carpenter, S. R., Roper, A. T., *A Guide Book for Technological Assessment and Impact Analysis*, New York: North-Holland, 1980.

Roberts, E. B., "Exploratory and Normative Technological Forecasting: A Critical Appraisal," *Technological Forecasting*, 1, 2, Fall 1969.

Thurston, P. H., "Make TF Serve Corporate Planning," *Harvard Business Review*, Sept–Oct. 1971.

Twiss, B. C., "Forecasting Market Size and Market Growth Rates for New Products," *Journal of Product Innovation Management*, 1, 1984, pp. 19–29.

Wheelwright, S. C. and Makridakis, S., *Forecasting Methods for Management*, New York: John Wiley & Sons, 1980.

Wise, G., "The Accuracy of Technological Forecasts 1890–1940," *Futures*, Oct. 1976, pp. 411–419.

1.2 Societal Values

Adler, P. S., "Automation and Skill: New Directions," *International Journal of Technology Management*, 2, 5–6, 1987, pp. 761–772.

Adler, P. S., "New Technology, New Skills," *California Management Review*, Fall, 1986, pp. 9–28.

Andrews, K. R., *The Concept of Corporate Strategy*, Homewood, Il: Richard D. Irwin, 1980.

Ansoff, H. I., *Implanting Strategic Management*, Englewood Cliffs, NJ: Prentice Hall International, 1984.

Braverman, H., *Labor and Monopoly Capital*, New York: Monthly Review Press, 1974.

Brown, J. S. and Newman, S. E., "Issues in Cognitive and Social Ergonomics: From Our House to Bauhaus," *Human-Computer Interaction*, 1, 1985, pp. 359–391.

Cockburn, C., *Brothers: Male Dominance and Technological Change*, London: Pluto Press, 1983.

Dickson, D., *Alternative Technology and the Politics of Technical Change*, London: Fontana/Collins, 1974, p. 224.

Ewing, D. W., *Do it My Way—or You're Fired*, New York: Wiley, 1983.

Fleischer, M., Liker, J. K. and Ansdorf, D. R., "Implementation and Use of Computer-aided Design and Computer-aided Engineering," *Transactions of 1987 IEEE Conference on Management of Evolving Systems*, Atlanta, Georgia.

Graham, J. W., "Principled Organizational Dissent: a Theoretical Essay," in L. L. Cummings and B. M. Staw, eds., *Research in Organizational Behavior*, 8, Aug 1986, pp. 1–52.

Harding, S. G., *The Science Question in Feminism*, Ithaca: Cornell University Press, 1986.

Hounshell, D. A., *From the American System to Mass Production, 1800–1932: The Development of Manufacturing Technology in the United States*, Baltimore, MD: Johns Hopkins University Press, 1984.

Howard, R., *Brave New Workplace*, New York: Viking, 1985.

Kelly, P. and Kranzberg, M., eds., *Technological Innovation: A Critical Review of Current Knowledge*, San Francisco: San Francisco Press, 1978.

Knights, D., Willnott, H. and Collinson, D., eds., *Job Redesign: Critical Perspectives on the Labour Process*, Hampshire, UK: Gower, 1985.

Knorr-Cetina, K. and Mulkay, M., eds., *Science Observed: Perspectives on the Social Study of Science*, London: Sage, 1983.

Krohn, W., Layton, E. T. and Weingart, P., eds., *The Dynamics of Science and Technology: Social Values, Technical Norms and Scientific Criteria in the Development of Knowledge*, Dordrecht, Holland: Reidel Publishing Company, 1978.

Latour, B. and Woolgar, S., *Laboratory Life: The Social Construction of Scientific Facts*, Beverly Hills: Sage, 1979.

Lauden, R., ed., *The Nature of Technological Knowledge: Are Models of Scientific Change Relevant?* Dordrecht: Reidel, 1984.

MacKenzie, D. and Wajcman, J., eds., *The Social Shaping of Technology*, Milton Keynes, UK: Open University Press, 1985.

Near, J. P. and Miceli, M. P., "Whistle-Blowers in Organizations: Dissidents or Reformers?," in L. L. Cummings and B. M. Staw, eds., *Research in Organizational Behavior*, 9, Sept 1987, pp. 321–368.

Noble, D. F., *America by Design*, New York: Knopf, 1977.

Noble, D. F., *Forces of Production*, New York, Oxford University Press, 1984.

Noble, D. F., "Social Choice in Machine Design: The Case of Automatically Controlled Machine Tools," in A. Zimbalist, ed., *Case Studies on the Labor Process*, New York: Monthly Review Press, 1979.

Piore, M. and Sabel, C. F., *The Second Industrial Divide,* New York: Basic Books, 1984.
Rogers, E. M., with Shoemaker, F. F., *Communication of Innovations: A Cross-Cultural Approach,* New York: Free Press, 1971.
Rosenberg, N. and Birdzell, L. E., Jr., *How the West Grew Rich,* New York: Basic Books, 1985.
Rothschild, J. ed., *Women, Technology and Innovation,* London: Pergamon, 1982.
Simmonds, W. H. C., "Impact of New External Values on R&D," *Research Management,* May 1978, pp. 29–33.
Smith, M. R., *Harpers Ferry Armory and the New Technology: The Challenge of Change,* Ithaca: Cornell University Press, 1977.
Weber, M., *The Protestant Ethic and the Spirit of Capitalism,* New York: Scribner, 1930.
Whalley, P., *The Social Production of Technical Work: The Case of British Engineers,* Albany: State University of New York Press, 1986.
Wood, S., ed., *The Degradation of Work? Skill, Deskilling and the Labour Process,* London: Hutchinson, 1982.
Zussman, R., *Mechanics of the Middle Class: Work and Politics Among American Engineers,* Berkeley: University of California Press, 1985.

1.3 Environmental Context: Government Policy

1.3.1 Federal Investment in R&D

Adams, *The Iron Triangle: The Politics of Defense Contracting,* New York: Council on Economic Priorities, 1981.
Barth, J., Cordes, J. and Tassey, G., "The Impact of Recent Changes in Tax Policy on Innovation and R&D," in B. Bozeman, M. Crow and A. N. Link, eds., *Strategic Management of Industrial R&D,* Lexington MA: Lexington, 1984.
Bozeman, B. and Link, A. N., "Tax Incentives for R&D: A Critical Evaluation," *Research Policy,* June 1984, pp. 21–31.
Bozeman, B. and Link, A. N., "Public Support for Private R&D: The Case of the Research Tax Credit," *Journal of Policy Analysis and Management,* Spring 1985, pp. 370–382.
Frame, J. D., "Tax Considerations in R&D Planning," *IEEE Transactions on Engineering Management,* EM-31, 2, May 1984.
Harvard Business School, "Note on the Aerospace Industry and Modernization," HBS Case #9-687-009, 1986.
Kaplan, R. S., *Tax Policies for R&D and Technological Innovation: Tax Policies of US and Foreign Nations in Support of R&D and Innovation,* Washington DC: National Technical Information Service, 1975.
Link, A. N., "The Impact of Federal Research and Development Spending on Productivity," *IEEE Transactions on Engineering Management,* EM-29, 4, Aug 1982, pp. 166–169.
Mansfield, E., "Federal Support of R&D Activities in the Private Sector," in *Priorities and Efficiency in Federal Research and Development,* Joint Economic Committee of Congress, 1976.
Mansfield, E., "Public Policy toward Industrial Innovation: An International Study of Direct Tax Incentives for Research and Development," in K. B. Clark, R. H. Hayes,

and C. Lorenz, eds., *The Uneasy Alliance: Managing the Productivity—Technology Dilemma,* Boston MA: Harvard Business School Press, 1985.

Mansfield, E., "The R&D Tax Credit and Other Technology Policy Issues," *American Economic Review,* May 1986, pp. 190–194.

Roberts, E. B., "Follow-On Contracts in Government-Sponsored Research and Development: Their Predictability and Impact," *Industrial Management Review,* 9, 2, 1968: 41–55.

Rosenberg, N., "*A Historical Overview of the Evolution of Federal Investment in R&D Since World War II,*" Paper commissioned for a workshop on The Federal Role in Research and Development, November 21–24, 1985, National Academy of Sciences, National Academy of Engineering, Institute of Medicine.

Souder, W. E. and Rubenstein, A. H., "Some Designs for Policy Experiments and Government Incentives for the R&D/Innovation Process," *IEEE Transactions on Engineering Management,* EM-23, 3, Aug 1976, pp. 129–139.

Tyler, P., *Running Critical: The Silent War, Rickover, and General Dynamics,* New York: Harper and Row, 1986.

1.3.2 Patents [see also 2.1.4.5a]

Alderson, W., Terpstra, V. and Shapiro, S. J., *Patents and Progress,* Homewood, IL: Richard D. Irwin, 1965.

Gilbert, R. J., "Patents, Sleeping Patents, and Entry Determents," in S. Slalop ed., *Strategy, Predation and Antitrust Analysis,* Washington, DC: Federal Trade Commission, 1981.

Griliches, Z., ed., *R&D, Patents, and Productivity,* National Bureau of Economic Research, University of Chicago Press, 1984.

Harbridge House Inc., *Government Patent Policy Study,* Washington, DC: U.S. Government Printing Office, 1968.

Kahn, A. E., "The Economics of Patent Policy," in J. P. Miller, ed., *Competition, Cartels and Their Regulation,* North-Holland, Amsterdam, 1962.

Kamien, M. I. and Schwartz, "Patent Life and R&D Rivalry," *American Economic Review,* 64, 1974, pp. 183–187.

Land, E., "Patents and New Enterprises," *Harvard Business Review,* 37, 5, Sept–Oct 1959.

Levin, R. C., Klevorick, A. K., Nelson, R. R. and Winter, S. G., "Appropriating the Returns from Industrial Research and Development," *Brookings Papers on Economic Activity,* 3, 1987, pp. 783–831.

Lieberman, M. B., "*Patents, Learning by Doing and Market Structure in the Chemical Processing Industries,*" Stanford University Graduate School of Business Working Paper No. 859, 1985.

Machlup, F., *The Economics of the Patent System,* Princeton University Press, 1953.

Mansfield, E., Schwatz, M. and Wagner, S., "Imitation Costs and Patents: An Empirical Study," *The Economic Journal,* 91, Dec 1981, pp. 907–918.

Nelson, R. R. ed., *The Rate and Direction of Inventive Activity,* National Bureau of Economic Research, Princeton University Press, 1962.

Preston, L., "Patent Rights Under Federal R&D Contracts," *Harvard Business Review,* 42, 5, Sep–Oct 1963.

Reich, L. S., "Research, Patents, and the Struggle to Control Radio: A study of Big Business and the Uses of Industrial Research," *Business History Review*, 1977.
Riggs, H. E., *Managing High-Technology Companies*, New York: Van Nostrand Reinhold, 1983.
Rothchild, R., "Making Patents Work for Small Companies," *Harvard Business Review*, 65, 4, Jul–Aug 1987.
Sanders, B., "Patent Utilization," *Patent, Trademark, and Copyright Journal*, Conference Supplement, 1957.
Scherer, F. M. et al., *Patents and the Corporation*, Harvard Business School, Boston, 1959.
Scherer, F. M., *Innovation and Growth*, Cambridge, MA: MIT Press, 1984.
Schmookler, J., *Invention and Economic Growth*, Cambridge, MA: Harvard University Press, 1966.
Spencer, R., "Threat to Our Patent System," *Harvard Business Review*, 34, 3, May–June 1956.
Taylor, C. T. and Silberston, Z. A., *The Economic Impact of the Patent System: A Study of the British Experience*, Cambridge, UK: Cambridge University Press, 1973.
Wyatt, S. and Bertini, G., *The Role of Patents in Multinational Corporations Strategies for Growth*, Paris: AREPIT, 1985.

1.3.3 Industrial/Technology Policy

Abernathy, W. J. and Chakravarthy, B. S., "Government Intervention and Innovation in Industry: A Policy Framework," *Sloan Management Review*, Spring 1979, pp. 3–18.
Allen, T. J., Utterback, J. M., Sirbu, M. S., Ashford, N. A. and Hollomon, J. H., "Government Influence on the Process of Innovation in Europe and Japan," *Research Policy*, 7, 2, April 1978.
Arthur, D. L., *Barriers to Innovation in Industry: Opportunities for Public Policy Changes, Main Report*, Washington DC: Arthur D. Little, 1973.
Bozeman, B. and Link, A. N., *Investments in Technology: Corporate Strategies and Public Policy Alternatives*, New York: Praeger, 1983.
Carter, C., ed., *Industrial Policy and Innovation*, London: Heinmann, 1981.
Center for Policy Alternatives, *National Support for Science and Technology: An Examination of Foreign Experience, 2 Volumes*, Cambridge, MA: Center for Policy Alternatives, MIT, 1975.
Commission on Government Procurement, *Report of the Commission on Government Procurement*, Washington, DC: GPO, 1972.
Dermer, J., ed., *Competitiveness Through Technology: What Business Needs from Government*, Lexington, MA: Lexington, 1986.
EIRMA, *Industry—Government Relations in R&D*, Paris: EIRMA, 1975.
Fusfeld, H. I., *The Technical Enterprise*, Cambridge, MA: Ballinger, 1986.
Gerstenfeld, A. and Brainard, R., eds., *Technological Innovation: Government/Industry Cooperation*, New York, John Wiley, 1979.
Holloway, J. H. and Assoc., "Government and the Innovation Process," *Technology Review*, May 1979, pp. 29–41.

Horwitch, M., *Clipped Wings,* Cambridge, MA: MIT Press, 1982.

Jarboe, K. P., ''A Reader's Guide to the Industrial Policy Debate,'' *California Management Review,* XXVII, 4, Summer 1985, pp. 198–219.

Kawase, T. and Rubenstein, A. H., ''Reactions of Japanese Industrial Managers to Government Incentives to Innovation—An Empirical Study,'' *IEEE Transactions on Engineering Management,* EM-24, 3, Aug 1977, pp. 93–101.

Kuehn, T. J. and Porter, A. L., eds., *Science, Technology and National Policy,* Ithaca: Cornell University Press, 1981.

Melman, S., *Profits Without Production,* New York: Knopf, 1983.

National Research Council, Office of Scientific and Engineering Personnel, Panel on Engineering Labor markets, *The Impact of Defense Spending on Nondefense Engineering Labor Markets: A Report to the National Academy of Engineering,* Washington, DC: National Academy Press 1986.

Nelson, R. R., Peck, M. and Kalachek, E., *Technology, Economic Growth, and Public Policy,* Brookings Institution, Washington, D.C., 1967.

Nelson, R. R. ed., *Government and Technical Change: A Cross-Industry Analysis,* NY: Pergamon Press, 1982.

Pavitt, K. and Walker, W., ''Government Policies Towards Industrial Innovation: A Review,'' *Research Policy,* 5, 1, 1976, pp. 11–97.

Peck, M. J. and Scherer, F. M., *The Weapons Acquisition Process: An Economic Analysis,* Boston: Harvard Business School, 1962.

Rothwell, R. and Zegveld, W., *Industrial Innovation and Public Policy: Preparing for the 1980s and the 1990s,* London: Francis Pinter, 1981, p. 251.

Schnee, J. R., ''Government Programs and the Growth of High-Technology Industries,'' *Research Policy,* 2, January 1978, pp. 2–24.

Tassey, G., ''Infratechnologies and the Role of Government,'' *Technological Forecasting and Social Change,* July 1982, 163–180.

Tassey, G., ''The Role of the National Bureau of Standards in Supporting Industrial Innovation,'' *IEEE Transactions on Engineering Management,* August 1986, pp. 162–171.

''The Defense Program and the Economy,'' *Hearings before the Subcommittee on Economic Goals and Intergovernmental Policy of Joint Economic Committee, U.S. Congress,* October 7, 13, 22, and 29, and December 15, 1982, Washington, DC, GPO 1983.

Utterback, J. M. and Murray, A. E., ''*The Influence of Defense Procurement and Sponsorship of Research and Development on the Development of the Civilian Electronics Industry,*'' MIT Center for Policy Analysis Working Paper CPA-77-2, June 1977.

Wachter, M. L. and Wachter, S. M. eds., *Towards a New U.S. Industrial Policy?* Philadelphia: University of Pennsylvania Press, 1983.

Yoshino, M. and Fong, G. R., ''The Very High Speed Integrated Circuit Programs: Lessons for Industrial Policy,'' in B. Scott and G. C. Lodge, eds., *US Competitiveness in the World Economy,* Boston, MA: Harvard Business School Press, 1985.

Zysman, J. and Tyson, L. eds., *American Industry in International Competition: Government Policies and Corporate Strategies,* Ithaca, NY: Cornell University Press, 1983.

1.3.4 Regulation [see also 2.1.4.8c]

Abernathy, W. J. and Chakravarthy, B. S., "Government Intervention and Innovation in Industry: A Policy Framework," *Sloan Management Review,* 20, 3, Spring 1979.

Antitrust Division, U.S. Department of Justice "Antitrust Guide for Joint Research Programs," *Research Management,* March 1981, pp. 30–37.

Bauer, R. A., de Sola Pool, I., and Dexter, L. A., *American Business and Public Policy,* New York: Atherton, 1964.

Baxter, W. F., "Antitrust Law and Technological Innovation," *Issues in Science and Technology,* Winter 1985, pp. 80–91.

Birnbaum, P. H., "The Choice of Strategic Alternatives Under Increasing Regulation in High Technology Companies," *Academy of Management Journal,* 27, 3, 1984, pp. 489–510.

Capron, W. M., *Technological Change in Regulated Industries,* Washington: The Brookings Institution, 1971.

Caves, R. E. and Roberts, M. J., eds., *Regulating the Product: Quality and Variety,* Cambridge, MA: Ballinger, 1975.

Chakrabarti, A. K. and Souder, W. E., "Government Regulations: Barriers or Stimuli to New Product Innovation?," *Contemporary Marketing Thought,* Chicago: American Marketing Association, 1977.

Ewing, K. P., Jr., "Joint Research, Antitrust and Innovation," *Research Management,* March–April 1981, pp. 25–29.

Hauptman, O. and Roberts, E. B., "FDA Regulation of Product Risk and its Impact Upon Young Biomedical Firms," *Journal of Product Innovation Management,* 4, 2, June 1987.

Joglekar, P. and Hamburg, M., "An Evolution of Federal Policies Concerning Joint Ventures for Applied Research and Development," *Management Science,* 29, 9, Sept. 1983, pp. 1016–1026.

Kitti, C. and Trozzo, C. L., *The Effects of Patent and Antitrust Laws, Regulations, and Practices on Innovation,* Virginia: Institute for Defense Analyses Program Analysis Division, 1976, 3, IDA Log No. HQ 76-18304.

National Academy of Sciences, *Medical Technology and the Health Care System: A Study of Diffusion of Equipment-Embodied Technology,* Washington, DC: National Research Council 1979.

Nelkin, D., ed, *Controversy: Politics of Technical Decision,* Beverly Hills, CA: Sage, 1984.

Ordover, J. A. and Willig, R., "Antitrust of High-Technology Industries: Assessing Research Joint Ventures and Mergers," *Journal of Law and Economics,* May 1985, pp. 311–333.

Otway, H. and Petty, M. eds., *Regulating Industrial Risks,* London: Butterworths, 1985.

Perrow, C., *Normal Accidents,* New York: Basic, 1984.

Roberts, E. B., "Influences on Innovation: Extrapolations to Biomedical Technology," in E. B. Roberts, R. I. Levy, S. N. Finkelstein, J. Moskowitz and E. J. Sondik eds, *Biomedical Innovation,* Cambridge, MA: MIT Press, 1981.

Six Countries Programme, *Innovation in Industry in Relation to Regulations in the Areas of Environmental Protection Occupational Health and Safety and in Telecommunica-*

tions. Papers Presented at the Joint Workshop of the Six Countries Programme and N.S.F., Washington, D.C., Held in The Hague, June 11–13, 1979.

Starr, C., "Industrial Cooperation in R&D," *Research Management*, Sept–Oct 1985, pp. 13–15.

Stigler, G. J., "The Theory of Economic Regulation," *Bell Journal of Economics and Management Science*, II, 1971, pp. 3–21.

Temin, P., "Technology, Regulations and Market Structure in the Modern Pharmaceutical Industry," *Bell Journal of Economics*, 10, 1979, pp. 426–446.

1.3.5 Education

Committee on the Education and Utilization of the Engineer, National Research Council, *Engineering Education and Practice in the United States: Foundations of Our Techno-Economic Future*, Washington, DC: National Academy Press, 1985.

Computerized Manufacturing Automation: Employment, Education and the Workplace, Washington, D.C., U.S. Congress OTA-CIT-235, 1984.

Francis, P. H., Phillips, W. M., DeVries, M. F., John, J. E. A., Goldhar, J. D. and Rivin, E. I., "The Academic Preparation of Manufacturing Engineers: A Blueprint for Change," *Manufacturing Review*, 1, 3, Oct 1988, pp. 158–163.

Nation Commission on Excellence in Education, *A Nation at Risk: The Imperative for Educational Reform*, Washington, DC: The Commission, 1983.

National Science Board, *Educating Americans for the 21st Century: A Report to the American People and the National Science Board*, Washington, DC: National Science Foundation 1984.

National Science Board, *Science and Engineering Indicators—1987*, Washington, DC: U.S.G.P.O. Office of Technology Assessment.

1.4 Demand

Clark, K. B., "The Interaction of Design Hierarchies and Market Concepts in Technological Evolution," *Research Policy*, 14, 1985, pp. 235–251.

Gold, B., "On the Adoption of Technological Innovations in Industry—Superficial Models and Complex Decision Processes," *Omega*, 8, 5, 1980, pp. 505–516.

Massy, W. F., Montgomery, D. B. and Morrison, D. G., *Stochastic Models of Buying Behavior*, Cambridge, MA: MIT Press, 1970.

Montgomery, D. and Urban, G. L., *Management Science in Marketing*, Englewood-Cliffs, NJ: Prentice-Hall, 1969.

Mowery, D. C. and Rosenberg, N., "The Influence of Market Demand upon Innovation: A Critical Review of Some Recent Empirical Studies," in N. Rosenberg, *Inside the Black Box: Technology and Economics*, Cambridge, UK: Cambridge University Press, 1982.

Robinson, D. E., "Style Changes: Cyclical, Inexorable and Foreseeable," *Harvard Business Review*, Nov–Dec. 1975.

Rogers, E. M., *Diffusion of Innovations*, New York: Free Press, 1983.

Rosenbloom, R. S. and Cusumano, M. A., "Technological Pioneering: The Birth of the

VCR Industry,'' *California Management Review,* XXIX, 4, Summer 1987, pp. 51–76.

Urban, G. L. and Hauser, J. R., *The Design and Marketing of New Products,* Englewood-Cliffs, NJ: Prentice-Hall, 1980.

Von Hippel, E. A., ''Successful Industrial Products from Customer Ideas,'' *Journal of Marketing,* January 1978.

Von Hippel, E. A., ''The Dominant Role of Users in the Scientific Instrument Innovation Process,'' *Research Policy,* 5, 1976, pp. 212–239.

Von Hippel, E. A., ''Users as Innovators,'' *Technology Review,* October–November 1977.

1.5 Competitive Behavior

1.5.1 Models

Arthur, W. B., ''*Competing Techniques and Lock-in by Historical Events: The Dynamics of Allocation Under Increasing Returns,*'' CEPR, Stanford University, 1985.

Barney, J. B., ''Types of Competition and the Theory of Strategy: Towards an Integrative Framework,'' *Academy of Management Review,* 11, 4, 1985, pp. 791–800.

Blaug, M., ''A Survey of the Theory of Process Innovation,'' *Economica,* 30, 1, 1963, pp. 13–32.

Caves, R. E., ''Industrial Organization, Corporate Strategy, and Structure,'' *Journal of Economic Literature,* 18, 1980, pp. 64–92.

Griliches, Z., ''Hybrid Corn: An Exploration in the Economics of Technological Change,'' *Econometrica,* Oct 1967, 25, 4, pp. 501–522.

Klein, B., *Dynamic Economics,* Cambridge, MA: Harvard Business University Press, 1977.

Leibenstein, H., ''Allocative Efficiency vs. X-Efficiency,'' *American Economic Review,* 56, 1966, pp. 392–415.

Mansfield, E., ''How Economists See R&D,'' *Harvard Business Review,* Vol. 59, No. 6, Nov–Dec 1981, pp. 98–106.

Mansfield, E., Rapoport, J., Schnee, J., Wagner, S. and Hamburger, M., *Research and Innovation in the Modern Corporation,* New York: Norton, 1971.

Mansfield, E., *The Production and Application of New Industrial Technology,* NY: Norton, 1977.

Nelson, R. R., ''Assessing Private Enterprise,'' *Bell Journal of Economic,* Spring 1981, pp. 93–111.

Nelson, R. R., ''Research on Productivity Growth and Productivity Differences: Dead Ends and New Departures,'' *Journal of Economic Literature,* Sept. 1981, pp. 1029–1064.

Nelson, R. R., *The Economics of Invention: A Survey of the Literature,* Santa Monica: Rand Corporation, 1980.

Nelson, R. R., ''The Simple Economics of Basic Scientific Research,'' *Journal of Political Economy,* 1959.

Pavitt, K., ''Technology, Innovation and Industrial Development: The New Causality,'' *Futures,* Dec. 1979, pp. 458–470.

Penrose, E. T., *The Theory of Growth of The Firm,* White Plains, NY: M. E. Sharpe, 1980.

Porter, M. E., "The Technological Dimension of Competitive Strategy," in R. S. Rosenbloom, ed., *Research on Technological Innovation, Management and Policy,* 1, 1983, pp. 1–33.

Porter, M. E., *Competitive Strategy,* New York: Free Press, 1980.

Porter, M. E., *Competitive Advantage,* New York: Free Press, 1985.

Rosenbloom, R. S. and Abernathy, W. J., "The Climate for Innovation in Industry: The Role of Management Attitudes and Practices in Consumer Electronics," *Research Policy,* 11, 1982, pp. 209–225.

Salter, W. G., *Productivity and Technical Change,* 2nd ed., Cambridge: Cambridge University Press, 1969.

Schmookler, J., *Invention and Economic Growth,* Cambridge: Harvard University Press, 1966.

Shumpeter, J. A., *The Theory of Economic Development,* Cambridge: Cambridge University Press, 1934.

Stiglitz, J. E., "Information and Economic Analysis: A Perspective," *Econ. J. Conference Papers,* 95, 1984, pp. 21–41.

Stoneman, P., *The Economic Analysis of Technological Change,* Oxford: Oxford University Press, 1983.

Winter, G., "Competition and Selection," in *The New Palgrave: A Dictionary of Economics,* London: MacMillan, 1987.

Winter, S. G., "Knowledge and Competence as Strategic Assets," in D. J. Teece, ed., *The Competitive Challenge,* Cambridge, MA: Ballinger, 1987.

1.5.2 International Comparisons

Abegglen, J. and Stalk, G., *Kaisha: The Japanese Corporation,* Basic Books, New York, 1985.

Bailey, M. N. and Chakrabarti, A. K., "Innovation and U.S. Competitiveness," *Brookings Review,* Fall 1985.

Chakrabarti, A. K., Feinman, S. and Fuentavilla, W., "Industrial Product Innovation: An International Comparison," *Industrial Marketing Management,* 7, 1978, pp. 231–237.

Clark, R., *Aspects of Japanese Commercial Innovation,* The Technical Change Center, London, Nov. 1984.

Fores, M., Sorge, A. and Lawrence, P., "Why Germany Produces Better," *Management Today,* November 1978.

Frame, J. D., *International Business and Global Technology,* Lexington, MA: Lexington, 1983.

Gold, B., "Factors Stimulating Technological Progress in Japanese Industries: The Case of Computerization in Steel," *Quarterly Review of Economics and Statistics,* Winter 1978, pp. 7–21.

Gold, B., "Technological and Other Determinants of the International Competitiveness of US Industries," *IEEE Transactions on Engineering Management,* EM-30, 2, May 1983, pp. 53–59.

Lynn, L., *How Japan Innovates,* Boulder, CO: Westview Press, 1982.
Mansfield, E., "The Speed and Cost of Industrial Innovation in Japan and the United States: External vs. Internal Technology," *Management Science,* 34, 10, Oct. 1988, pp. 1157–1168.
Nelson, R. R., *High Technology Policies: A Five-Nation Comparison,* Washington, DC: American Enterprise Institute, 1984.
Piore, M. J. and Sabel, C. J., *The Second Industrial Divide,* New York: Basic Books, 1984.
Roman, D. D. and Puett, J. F., Jr., *International Business and Technological Innovation,* New York: North Holland, 1983.
Rothwell, R. and Wissema, H., "Technology, Culture and Public Policy," *Technovation,* 4, 1985, pp. 91–115.
Scott, B. R. and Lodge, G. C., eds., *US Competitiveness in The World Economy,* Boston, MA: Harvard Business School Press, 1985.
Stobaugh, R. and Wells, Jr. L. T., eds., *Technology Crossing Borders,* Boston, MA: Harvard Business School Press, 1984.
Tsuchiya, M., "*A Bird's-Eye View of the Technological Innovation in Japanese Business,*" Paper presented at the Mitsubishi Bank Foundation Conference on Business Strategy and Technical Innovation, March 1985, Itoh City, Japan.
Utterback, J. M., Meyer, M., Roberts, E. B. and Reitberger, G. "Technology and Industrial Innovation in Sweden: A Study of Technology-based Firms Formed Between 1965 and 1980," *Research Policy,* 17, 1988, pp. 15–26.
Westney, D. E. and Sakakibara, K., "The Challenge of Japan-Based R&D in Global Technology Strategy," in M. Horwitch, ed., *Technology in Society,* 7, 2/3, 1986.

1.5.3 Innovation and Market Structure

Cohen, W. M. and Levin, R. C., "Empirical Studies of Innovation and Market Structure," in R. Schmalensee and R. Willig, eds., *The North-Holland Handbook of Industrial Organization,* New York: North-Holland, 1988, forthcoming.
Cohen, W. M. and Mowery, D. C., "Firm Heterogeneity and R&D: An Agenda for Research" in B. Bozeman, M. Crow and A. Link, eds., *Strategic Management of Industrial R&D,* Lexington, MA: D. C. Health, 1984.
Flaherty, M. T., "Technological Leadership and Market Share Determination," Harvard Business School, 1984.
Kamien, M. I. and Schwartz, N. L., *Market Structure and Innovation,* Cambridge, MA: Cambridge University Press, 1982.
Kelly, P. and Kranzberg, M., eds., *Technological Innovation: A Critical Review of Current Knowledge,* San Francisco: San Francisco Press, 1978.
Levin, R. C., Klevorick, A. K., Nelson, R. R. and Winter, S. G., "*Survey Research on R&D Appropriability and Technological Opportunity. Part 1: Appropriability,*" Yale University, July 1984.
Mowery, D. C., "Innovation, Market Structure, and Government Policy in the American Semiconductor Industry: A Survey," *Research Policy,* 12, 4, Aug 1983.
Nelson, R. R. and Winter, S. G., *An Evolutionary Theory of Economic Change,* Cambridge, MA: Harvard University Press, 1982.
Soete, L., "Firm Size and Inventive Activity: The Evidence Reconsidered," *European Economic Review,* Vol. 12, 1979, pp. 319–340.

Stoneman, P., *The Economic Analysis of Technological Change,* Oxford: Oxford University Press, 1983.

Sylos-Labini, P., *Oligopoly and Technical Progress,* 2nd ed., Cambridge University Press, 1967.

1.5.4 Vertical Integration [see also 2.1.4.5a]

Armour, H. W. and Teece, D. J., "Vertical Integration and Technological Innovation," *Review of Economics and Statistics,* 62, 1980, pp. 470–474.

Balakrishnan, S. and Wernerfelt, B., "Technical Change, Competition and Vertical Integration," *Strategic Management Journal,* 7, 1986, pp. 346–360.

Eccles, R. G., "The Quasi-Firm in the Construction Industry," *Journal of Economic Behavior and Organization,* 2, 1981, pp. 335–357.

Harrigan, K. R., "A Framework for Looking at Vertical Integration," *Journal of Business Strategy,* 3, 1983, pp. 30–37.

Harrigan, K. R., *Strategies for Vertical Integration,* Lexington: DC Health, 1983.

Monteverde, K. and Teece, D. J., "Supplier Switching Costs and Vertical Integration in the Automobile Industry," *Bell Journal of Economics,* 12, 1982, pp. 206–213.

Teece, D. J., "Profiting from Technological Innovation: Implications for Integration, Collaboration, Licensing and Public Policy," in D. J. Teece, ed., *The Competitive Challenge,* Cambridge, MA: Ballinger 1987.

1.5.5 Standards [see also 2.1.4.5a]

Barnett, W. P., "*The Organizational Ecology of the Early American Telephone Industry: A Study of the Technological Cases of Competition and Mutualism,*" Ph D. Dissertation, University of California, Berkeley, 1988.

David, P. A., "*Narrow Windows, Blind Giants and Angry Orphans: Dynamics of Systems Rivalries and Dilemmas of Technology Policy,*" Stanford Center for Economic Policy Research, Working Paper 10, 1986.

David, P. A., "Clio and the Economics of QWERTY," *American Economic Review,* 75, 2, May 1985, pp. 332–337.

David, P. A., "*Some New Standards for the Economics of Standardization in the Information Age,*" Stanford, Stanford Center for Economic Policy Research, Working paper 11, 1986.

Farrell, J. and Saloner, G., "*Economic Issues in Standardization,*" Department of Economics, MIT, Working Paper 393.

Farrell, J. and Saloner, G., "Standardization, Compatibility and Innovation," *Rand Journal,* 16, Spring 1985, pp. 70–83.

Gerstenfeld, A., "Government Regulation Effects on the Direction of Innovation: A Focus on Performance Standards," *IEEE Transactions on Engineering Management,* EM-24-3, Aug 1977, pp. 82–86.

Katz, M. L. and Shapiro, C., "Network Externalities, Competition and Compatibility," *American Economic Review,* 75, May 1985, pp. 424–440.

Keeney, R. L., "Issues in Evaluating Standards," *Interfaces,* 13, 2, 1983, pp. 12–22.

Kindleberger, C. P., "Standards as Public, Collective and Private Goods," *Kyklos,* 36, Fasc. 3, 1983, pp. 377–396.

2. ORGANIZATION

2.1 Technology Strategy Content

2.1.1 Business Strategy and Technology Strategy [see also 2.1.5.1]

Ansoff, H. I. and Stewart, J. M., "Strategies for a Technology-Based Business," *Harvard Business Review,* November–December 1967.

Bitondo, D. and Frohman, A., "Linking Technological and Business Planning," *Research Management,* Nov. 1981, pp. 19–23.

Burgelman, R. A., "*Strategy-Making and Evolutionary Theory: Towards a Capabilities-Based Perspective,*" Research Paper, Series 755, Graduate School of Business, Stanford University.

Cannon, P., "Integrative Planning and Communication of Research," *Research Management,* May–June 1984, pp. 20–23.

Frohman, A. L., "Putting Technology Into Strategic Planning," *California Management Review,* XXVII, 2, Winter 1985.

Harris, J. M., Shaw, R. W. Jr., Sommers, W. P., "The Strategic Management of Technology," *Booz, Allen & Hamilton: Outlook,* Fall/Winter 1981.

Hayes, R. H. and Wheelwright, S. C. *Restoring Our Competitive Edge: Competing Through Manufacturing,* New York: John Wiley & Sons, 1984, Chapter 13.

Hayes, R. H., "Strategic Planning: Forward in Reverse?" *Harvard Business Review,* November–December 1985.

Liberatore, M. J. and Titus, G. J., "*The Linkages Between R&D and Strategic Planning,*" Villanova University, Oct. 1987.

SEST-EUROCONSULT, "*Le bonzai de l'industrie japonaise.* Eléments de réflexion sur l'intégration de la technologie dans la function stratégique des entreprises japonaises," French Ministry of Research and Technology, July 1984.

Thomas, L. J., "Technology and Business Strategy—the R&D Link," *Research Management,* May–June 1984, pp. 15–19.

Weil, E. D. and Cangemi, R. R., "Linking Long-Range Research to Strategic Planning," *Research Management,* May–June 1983, pp. 32–39.

2.1.2 Types of Technology Strategy Models

Ansoff, H. I., *Implanting Strategic Management,* Englewood Cliffs, NJ: Prentice-Hall International, 1984.

Booz-Allen and Hamilton, "The Strategic Management of Technology," *Booz-Allen and Hamilton: Outlook,* Fall–Winter 1981.

Fusfield, A. R., "How to Put Technology into Corporate Planning," *Technology Review,* May 1978.

Gluck, F. W., Kaufman, S. P. and Walleck, A. S., "Strategic Management for Competitive Advantage," *Harvard Business Review,* July–August 1980.

Haspelagh, P., "Portfolio Planning: Uses and Limits," *Harvard Business Review,* Jan–Feb 1982.

Horwitch, M. and Pralahad, C. K., "Technological Innovation—Three Ideal Modes," *Sloan Management Review,* Winter 1976, pp. 77–89.

Porter, M. E., *Competitive Advantage,* New York: Free Press, 1985.

Quinn, J. B., *The Strategy Process,* Englewood Cliffs, NJ: Prentice-Hall, 1988.

Quinn, J. B., "Innovation and Corporate Strategy: Managed Chaos," in M. Horwitch, ed., *Technology in Society*, 2/3, 1986, pp. 167–183.

Rosenbloom, R. S., "Managing Technology for the Longer Term: A Managerial Perspective" in K. B. Clark, R. H. Hayes and C. Lorenz, eds., *The Uneasy Alliance: Managing the Productivity-Technology Dilemma*, Boston, MA: Harvard Business School Press, 1985.

2.1.3 *Identifying Mission-Objectives-Strategy*

A. D. Little, "The Strategic Management of Technology," *European Management Forum*, 1981.

Abell, D., *Defining the Business: The Starting Point of Strategic Planning*, Englewood Cliffs, N.J.: Prentice-Hall, 1980.

Abernathy, W. J., *The Productivity Dilemma: Roadblocks to Innovation in the Automobile Industry*, Baltimore: Johns Hopkins University Press, 1978.

Allen, J. W., "Research and Development: Managing for Productivity in the 1980's," *Booz, Allen & Hamilton: Outlook*, Fall/Winter 1980.

Anderson, C. R. and Paine, F. T., "PIMS: A Re-Examination," *Academy of Management Review*, 3, 1978, pp. 602–612.

Ansoff, H. I. and Stewart, J. M., "Strategies for a Technology-Based Business," *Harvard Business Review*, November–December 1967.

Ansoff, H. I., *Implanting Strategic Management*, Englewood Cliffs, NJ: Prentice-Hall International, 1984.

Baker, R. N., Green, S. G. and Bean, A. S., "The Need for Strategic Balance in R&D Project Portfolios," *Research Management*, March–April 1986, pp. 38–43.

Betz, F., *Managing Technology*, Englewood Cliffs, NJ: Prentice-Hall, 1987.

Birnbaum, P. H., "Strategic Management of Industrial Technology: A Review of the Issues," *IEEE Transactions on Engineering Management*, EM-31, 4, Nov 1984, pp. 186–191.

Bitondo, D., "Technology Planning in Industry: The Classical Approach" in M. J. Dluhy and K. Chen, eds., *Interdisciplinary Planning: A Perspective for the Future*, New Jersey: Center for Urban Policy Research, Rutgers University, 1986.

Boddy, D. and Buchanan, D. A., *Managing New Technology*, Oxford: Basil Blackwell, 1986.

Booz-Allen and Hamilton, "The Strategic Management of Technology," *Booz-Allen & Hamilton: Outlook*, Fall–Winter 1981.

Burgelman, R. A., Kosnik, T. J., Van den Poel, M., "Toward an Innovative Capabilities Audit Framework," in R. A. Burgelman, and M. A. Maidique, eds, *Strategic Management of Technology and Innovation*, Homewood, IL: Irwin, 1988.

Burns, T. and Stalker, G. M., *The Management of Innovation*, London: Tavistock, 1961.

"Chaparral Steel," Harvard Business School case 9-687-045, 1987.

Clark, K. B. and Griliches, Z., "Productivity Growth and R&D at the Business Level: Results from the PIMS Data Base," in Z. Griliches, ed., *R&D, Patents and Productivity*, Chicago: IL: University of Chicago Press, 1984.

Clark, K. B. and Hayes, R. H., "Exploring Factors Affecting Innovation and Productivity Growth Within the Business Unit," in K. B. Clark, R. H. Hayes and C. Lorenz, eds., *The Uneasy Alliance: Managing the Productivity—Technology Dilemma*, Boston: Harvard Business School Press, 1985.

Cooper, R. G. and Schendel, D., "Strategic Responses to Technological Threats," *Business Horizons,* 19, 1, Feb. 1976.

Cooper, R. G., "Defining the New Product Strategy," *IEEE Transactions on Engineering Management,* EM-34, Aug. 1987, pp. 184–193.

Cooper, R. G., "How New Product Strategies Impact on Performance," *Journal of Product Innovation Management,* January 1984.

Durand, T., "R&D Program-Competence Matrix," *R&D Management,* Feb. 1988.

Ettlie, J. E. and Bridges, W. P., "Environmental Uncertainty and Organizational Technology Policy," *IEEE Transactions on Engineering Management,* EM-29, 1, Feb. 1982, pp. 2–10.

Ettlie, J. E., Bridges, W. P. and O'Keefe, R. D., "Organization Strategy and Structural Differences for Radical Versus Incremental Innovation," *Management Science,* 30, 6, June 1984, pp. 682–695.

Fischer, W. A., "Follow-Up Strategies for Technological Growth," *California Management Review,* Fall 1978.

Fischer, W. A., "Follow-Up Strategies for Technological Threats," *Business Horizons,* February 1977.

Foster, R. N., *Innovation: The Attacker's Advantage,* New York: Summit, 1986.

Freeman, C. *The Economics of Industrial Innovation,* 2nd ed., Cambridge MA: MIT Press, 1986.

Frohman, A. L. and Bitondo, D., "Coordinating Business Strategy and Technical Planning," *Long Range Planning,* 14, Dec. 1981, pp. 58–67.

Frohman, A. L., "Managing the Company's Technological Assets," *Research Management,* Sept. 1980, pp. 20–24.

Frohman, A. L., "Putting Technology Into Strategic Planning," *California Management Review,* Winter 1985.

Gabel, L. ed., *Product Standardization and Competitive Strategy,* Amsterdam: North Holland, 1987.

Galbraith, J. R. and Kazanjian, R. K., "Developing Technologies: R&D Strategies of Office Products Firms," *Columbia Journal of World Business,* Spring 1983, pp. 37–44.

Goldring, L. S., "Defensive/Offense Research Synergy," *Research Management,* March 1979, pp. 30–32.

Hambrick, D. C. and MacMillan, I. C., "Efficiency of Product R&D in Business Units: The Role of Strategic Context," *Academy of Management Journal,* 28, 3, 1985, pp. 527–547.

Hariharan, S. and Prahalad, C. K., "*Technological Compatibility Choices in High-Tech Products: Implications for Corporate Strategy,*" Paper presented at the First Annual Conference on Managing the High-Tech Firm, Boulder, Colorado, Jan. 1988.

Harris, J. M., Shaw, R. W., Jr. and Sommers, W. P., "The Strategic Management of Technology," *Booz-Allen & Hamilton: Outlook,* Fall/Winter, 1981.

Hax, A. C. and Majluf, N. S., *Strategic Management: An Integrated Perspective,* Englewood Cliffs, NJ: Prentice-Hall, 1984.

Horwitch, M. and Pralahad, C. K., "Technological Innovation: Three Ideal Modes," *Sloan Management Review,* Winter 1976, pp. 77–89.

Horwitch, M. and Thiefort, R. A. "The Effect of Business Interdependencies on Product R&D-intensive Business Performance," *Management Science,* February 1987, pp. 178–197.

Iyer, E. S. and Ramaprasad, A., "Strategic Postures Toward Innovation," *IEEE Transactions on Engineering Management,* EM-31, 2, May 1984, pp. 87–90.

Johnson, S. B. and Jones, C., "How to Organize for New Products," *Harvard Business Review,* May–June 1957.

Johnson, S. B., "Comparing R&D Strategies of Japanese and U.S. Firms," *Sloan Management Review,* 25, 3, Spring 1984.

Ketterington, J. and White, J., "Making Technology Work for Business," in R. Lamb, ed., *Competitive Strategic Management,* Englewood Cliffs, NJ: Prentice-Hall, 1984.

Kiel, G., "Technology and Marketing: The Magic Mix?," *Business Horizons,* May–June, 1984.

Link, A. N. and Tassey, G., *Strategies for Technology-Based Competition,* Lexington, MA: Lexington, 1987.

Maidique, M. A. and Hayes, R. H., "The Art of High-Technology Management," *Sloan Management Review,* 25, Winter 1984, pp. 18–31.

Maidique, M. A. and Patch, P., "*Corporate Strategy and Technological Policy,*" Harvard Business School, Note 9-679-033, Rev. 3/80.

Mansfield, E., "Technical Change and the Rate of Imitation," *Econometrica,* 29, 1961.

Martin, M. J. C., *Managing Technological Innovation and Entrepreneurship,* Reston, VT: Reston, 1984.

Meredith, J., "The Role of Manufacturing Technology in Competitiveness: Peerless Laser Processors," *IEEE Transactions on Engineering Management,* EM-35, 1, Feb. 1988, pp. 3–10.

Meyer, M. H. and Roberts, E. B., "Focusing Product Technology for Corporate Growth," *Sloan Management Review,* Summer 1988.

Meyer, M. H. and Roberts, E. B., "New Product Strategy in Small Technology-Based Firms: A Pilot Study," *Management Science,* 37, 7, July 1986, pp. 806–821.

Mitchell, G. R. and Hamilton, W. F., "Managing R&D as a Strategic Option," *Research and Technology Management,* May–June 1988, pp. 15–22.

Mitchell, G. R., "New Approaches for the Strategic Management of Technology," in M. Horwitch, ed., *Technology in Society,* 7, 2/3, 1986, pp. 132–144.

Mowery, D. C. and Rosenberg, N., "The Influence of Market Demand Upon Innovation: A Critical Review of Some Resent Empirical Studies," in N. Rosenberg, *Inside The Black Box: Technology and Economics,* Cambridge, UK: Cambridge University Press, 1982.

Normann, R., "Organizational Innovativeness: Product Variation and Reorientation," *Administrative Science Quarterly,* 16, 1971, pp. 203–215.

Pascale, R. T. and Athos, A., *The Art of Japanese Management,* New York: Simon and Schuster, 1981.

Peirce, W. S. and Rosegger, G., "Diffusion of Major Technological Innovations in U.S. Iron and Steel Manufacturing," *Journal of Industrial Economics,* 18, 1970, pp. 219–222.

Peters, T. J. and Waterman, R. H., Jr., *In Search of Excellence,* New York: Harper and Row, 1982.

Petrov, B., "The Advent of the Technology Portfolio," *Journal of Business Strategy,* Fall 1982.

Porter, M. E., *Competitive Advantage,* New York: Free Press, 1985.

Quinn, J. B., "Innovation and Corporate Strategy: Managed Chaos," in M. Horwitch, ed., *Technology in Society,* 7, 2/3, 1986, pp. 167–183.

Quinn, J. B., "Managing Innovation: Controlled Chaos," *Harvard Business Review,* May–June 1985.
Riggs, H. E., *Managing High-Technology Companies,* New York: Van Nostrand Reinhold, 1983.
Roberts, E. B., "Technology Strategy for the Medium-Sized Company," *Research Management,* July 1976.
Rosenberg, N., *Inside the Black Box: Technology and Economics,* Cambridge, MA: Cambridge University Press, 1982.
Rosenbloom, R. S. and Cusumano, M. A., "Technological Pioneering: The Birth of the VCR Industry," *California Management Review,* XXIX, 4, Summer 1987, pp. 51–76.
Rosenbloom, R. S. and Wolek, F. W., *Technology and Information Transfer,* Boston, MA: Harvard University Graduate School of Business Administration, 1970.
Rosenbloom, R. S., "Managing Technology for the Longer Term: A Managerial Perspective" in K. B. Clark, R. H. Hayes and C. Lorenz, eds., *The Uneasy Alliance: Managing the Productivity-Technology Dilemma,* Boston: Harvard Business School Press, 1985.
Rothberg, R. R., Ed., *Corporate Strategy and Product Innovation,* 2nd ed., New York: Free Press, 1981.
Rothwell, R., "Technical Change and Competitiveness in Engineering Goods: Some European Data," *IEEE Transactions on Engineering Management,* EM-28, 4, Nov 1981, pp. 89–96.
Rumelt, R., *Strategy, Structure and Economic Performance,* Boston, MA: Division of Research, Harvard Business School, 1974.
Schein, E. H., "Coming to a New Awareness of Organizational Culture," *Sloan Management Review,* Winter 1984, pp. 3–16.
Snow, C. C. and Hbreniak, L. G., "Strategy, Distinctive Competence, and Organizational Performance," *Administrative Science Quarterly,* 25, June 1980.
Stevenson, H. H., "Defining Strengths and Weaknesses," *Sloan Management Review,* Spring, 1976.
Tucker, S. A., ed., *A Modern Design for Defense Decision: A McNamara-Hitch-Enthoven Anthology,* Washington, DC: Industrial College of the Armed Forces, 1966.
Tushman, M. L. and Anderson, P., "Technological Discontinuities and Organizational Environments," *Administrative Science Quarterly,* 31, 1986, pp. 439–465.
Twiss, B. C., *Managing Technological Innovation,* London: Longman, 1980.
Walton, R. E., "Challenges in the Management of Technology and Labor Relations," in R. E. Walton and P. R. Lawrence, eds., *HRM Trends and Challenges,* Boston, MA: Harvard Business School Press, 1985.
Wyman, J., "Technological Myopia: The Need to Think Strategically About Technology," *Sloan Management Review,* Summer 1985, pp. 59–64.

2.1.4 Operationalizing Technology Strategy: Policies

Ansoff, H. I. and Stewart, J. M., "Strategies for a Technology-Based Business," *Harvard Business Review,* November–December 1967.
Betz, F., *Managing Technology,* Englewood Cliffs, NJ: Prentice-Hall, 1987.
Birnbaum, P. A., "Strategic Management of Industrial Technology—A Review of the Issues," *IEEE Transactions on Engineering Management,* EM-31, 4, 1984.

Booz-Allen and Hamilton, "The Strategic Management of Technology," *Booz-Allen & Hamilton: Outlook,* Fall–Winter 1981.

Burgelman, R. A., Kosnik, T. J., Van den Poel, M., "Toward an Innovative Capabilities Audit Framework," in R. A. Burgelan, and M. A. Maidique, eds., *Strategic Management of Technology and Innovation,* Homewood, IL: Irwin, 1988.

Foster, R. N., *Innovation: The Attacker's Advantage,* New York: Summit, 1986.

Frohman, A. L., "Putting Technology Into Strategic Planning," *California Management Review,* XXXVII, 2, Winter 1985.

Galbraith, J. R. and Kazanjian, R. K., "Developing Technologies: R&D Strategies of Office Products Firms," *Columbia Journal of World Business,* Spring 1983, pp. 37–44.

Gerstenfeld, A., *Effective Management of Research and Development,* Reading, MA: Addison-Wesley, 1970.

Gobeli, D. H. and Rudelis, W., "Managing Innovation: Lessons From the Cardiac Pacing Industry," *Sloan Management Review,* Summer 1985, pp. 29–43.

Hayes, R. H., Wheelwright, S. C. and Clark, K. B., *Dynamic Manufacturing,* New York: Wiley, 1988.

Hayes, R. H. and Wheelwright, S. C., *Restoring Our Competitive Edge: Competing Through Manufacturing,* New York: John Wiley & Sons, 1984, Chapter 10.

Horwitch, M. and Pralahad, C. K., "Technological Innovation—Three Ideal Modes," *Sloan Management Review,* Winter 1976, pp. 77–89.

Liker, J. K. and Hancock, W. M., "Organizational Barriers to Engineering Effectiveness," *IEEE Transactions on Engineering Management,* EM-33, 2, May 1986, pp. 82–91.

Maidique, M. A. and Patch, P., "*Corporate Strategy and Technological Policy,*" Harvard Business School, Case #9-679-033, Rev. 3/80.

Martin, M. J. C., *Managing Technological Innovation and Entrepreneurship,* Reston, VT: Reston, 1984.

Mitchell, G. R. and Hamilton, W. F., "Managing R&D as a strategic Option," *Research and Technology Management,* May–June 1988, pp. 15–22.

Porter, M. E., *Competitive Advantage,* New York: Free Press, 1985.

Riggs, H. E., *Managing High-Technology Companies,* New York: Van Nostrand Reinhold, 1983.

Sen, F. and Chakrabarti, A., "*Factors Affecting the Formulation of Technology Strategies in High-Technology Firms.*" Paper presented at ORSA/TIMS Meetings, St. Louis, October 1987.

Tushman, M. L. and Moore, W. L. eds., *Readings on the Management of Innovation,* Boston: Pitman, 1982.

Twiss, B. C., *Managing Technological Innovation,* London: Longman, 1980.

2.1.4.1 Technical Personnel

2.1.4.1a Skills

Badawy, M. K., *Developing Managerial Skills in Engineers and Scientists,* New York: Van Nostrand Reinhold, 1982.

Bailyn, L., *Living with Technology: Issues at Mid-Career,* Cambridge, MA: MIT Press, 1980.

Buchanan, D. A. and Boddy, D., *Organizations in the Computer Age: Technological Imperatives and Strategic Choice,* Aldershot: Gower, 1983.

Cheng, J. L. C., "Organizational Staffing and Productivity in Basic and Applied Research: A Comparative Study," *IEEE Transactions on Engineering Management,* EM-31, 1, February 1984.

Dalton, G. W. and Thompson, P. H., "Accelerating Obsolescence of Older Engineers," *Harvard Business Review,* Sept–Oct., 1971.

Decker, W. D. and Van Atta, C. M., "Controlling Staff Age and Flexible Retirement Plans," *Research Management,* 1, 1973.

Denis, H., "Matrix Structures, Quality of Working Life, and Engineering Productivity," *IEEE Transactions on Engineering Management,* EM-33, 3, Aug. 1986, pp. 148–156.

Durand, T., "R&D Program-Competence Matrix," *R&D Management,* Feb. 1988.

Fraker, J. R., "Professional Training for the Engineer-Manager (Don't Forget the Engineering)," *IEEE Transactions on Engineering Management,* EM-27, 4, Nov 1980, pp. 109–110.

Goldner, F. H. and Ritti, R. R., "Professionalism as Career Immobility," *American Journal of Sociology,* 72, pp. 489–502.

Hall, D. and Lawler, E., III., "Unused Potential in R&D Organizations," *Research Management,* XII, 5, 1969, pp. 339–354.

Heimer, C. A., "Organizational and Individual Control of Career Development in Engineering Project Work," in A. L. Stinchcombe and C. A. Heimer, *Organization Theory and Project Management,* Oslo: Norwegian University Press, 1985.

Hurt, N. H., Russell, W. G., Davis, P. I., McGuire, G. E. and Nickel, D. L., "Educating the Engineering Manager," *IEEE 1986 World Conference on Continuing Engineering Education.*

Katz, R. and Tushman, M. L., "A Longitudinal Study of the Effects of Boundary Spanning Supervision on Turnover and Promotion in Research and Development," *Academy of Management Journal,* 26, 3, 1983, pp. 437–456.

Katz, R.; "Managing Careers: The Influence of Job and Group Longevities," in R. Katz, ed., *Career Issues in Human Resource Management,* Englewood Cliffs, N.J.: Prentice Hall, 1982.

Kaufman, H. G., "Continuing Education for Updating Technical People," *Research Management,* July 1975, pp. 20–23. 214.2

Keller, R. T. and Holland, W. E., "Toward A Selection Battery for Research and Development Professional Employees," *IEEE Transactions on Engineering Management,* EM-26, 4, Nov 1979, pp. 90–93.

Kerr, S. and Von Glinow, M. A., "Issues in the Study of Professionals in Organizations: The Case of Scientists and Engineers," *Organizational Behavior and Human Performance,* 18, 1977, pp. 329–345.

Kimblin, C. W. and Souder, W. E., "Maintaining Productivity as Staff Half-Life Decreases," *Research Management,* Nov. 1975, pp. 29–35.

Kirkham, K. and Thompson, P., "Managing a Diverse Workforce: Women in Engineering," *Research Management,* March–April 1984, pp. 9–16.

Kleingartner, A. and Anderson, C., *Human Resource Management in High-Technology Firms,* Lexington, MA: Lexington Books, 1987.

Landy, F. J., "Developing Scales for Measuring Obsolescence," *Research Management,* July 1975, pp. 11–14.

Lazer, R. I., "The Pros and Cons of Flexible Working Hours in R&D Labs," *Research Management,* Jan. 1980. pp. 19–22.

Lee, D. M. S., "Academic Achievement, Task Characteristics, and First Job Performance of Young Engineers," *IEEE Transactions on Engineering Management,* EM-33, 3, Aug 1986, pp. 127–133.

Maidique, M. A., "Entrepreneurs, Champions and Technological Innovation," *Sloan Management Review,* 21, 2, Winter 1980, pp. 59–76.

Majchrzak, A., "A National Probability Survey on Education and Training for CAD/CAM," *IEEE Transactions on Engineering Management,* EM-33, 4, Nov 1986, pp. 197–206.

Martino, J. P., "A Survey of Behavioral Science Contributions to Laboratory Management," *IEEE Transactions on Engineering Management,* EM-20, 3, Aug. 1973, pp. 68–75.

Miller, D. B., "Managing for Long-Term Technical Vitality," *Research Management,* July 1975, pp. 15–19.

Morton, J. A., *Organizing for Innovation,* New York: McGraw-Hill, 1971.

Orth, C. D., 3rd, "More Productivity from Engineers," *Harvard Business Review,* March–April 1957.

Pegels, C. C., "A Markov Chain Application to an Engineering Manpower Policy Problem," *IEEE Transactions on Engineering Management,* EM-28, 2, May 1981, pp. 39–42.

Phelps, E. D. and Gallagher, W., "Integrated Approach to Technical Staffing," *Harvard Business Review,* July–Aug 1963.

Pyler, W., "Accounting for People," in *Managing Advanced Technology, Vol. 2: Creating an Action Team in R&D,* New York: American Management Association, 1972.

Roberts, E. B. and Fusfeld, A. R., "Staffing the Innovative Technology-Based Organization," *Sloan Management Review,* 22, 3, Spring 1981.

Rosen, N., Billings, R. and Turney, J., "The Emergence and Allocation of Leadership Resources Over Time in a Technical Organization," *Academy of Management Journal,* vol. 19, 1976, pp. 165–183.

Shapero, A., "Managing Creative Professionals," *Research Management,* March–April 1985, pp. 23–28.

Swords-Isherwood, N. and Senker, P., eds., *Microelectronics and the Engineering Industry: The Need for Skills,* London: Francis Pinter, 1980.

Thompson, P. H. and Dalton, G. W., "Are R&D Organizations Obsolete?," *Harvard Business Review,* Nov–Dec. 1976.

Westney, E. and Sakakibara, K., "*Comparative Study of the Training, Careers and Organization of Engineers in the Computer Industry in Japan and the United States,*" MIT-Japan Science and Technology Program, Sept. 1985.

Whalley, P., *The Social Production of Technical Work: The Case of British Engineers,* Albany: State University of New York Press, 1986.

Wolff, M. F., "Hiring People Who Do Good Research," *Research Management,* Jan–Feb. 1984, pp. 8–9.

2.1.4.1b Evaluation and rewards

Allen, T. J. and Katz, R., "The Dual Ladder—Problem or Solution?" *IEEE Careers Conference,* 1985, pp. 142–146.

Badawy, M. K., "Applying Management by Objectives to R&D Labs," *Research Management,* Nov. 1976, pp. 35–40.
Badawy, M. K., "Industrial Scientists and Engineers: Motivational Style Differences," *California Management Review,* Fall 1971, pp. 11–16.
Badawy, M. K., "One More Time: How to Motivate Your Engineers," *IEEE Transactions on Engineering Management,* EM-25, 2, May 1978, pp. 37–42.
Balkin, D. B. and Gomez-Mejia, L. R., "Determinants of R and D Compensation Strategies in High Tech Industries," *Personnel Psychology,* 37, 1984, pp. 635–650.
Bristol, R. O., "A Look at Employee Fringe Benefits as a Means of Partial Compensation," *IEEE Transactions on Engineering Management,* EM-28, 2, May 1981, pp. 26–30.
Cantrall, E. W., "The Dual Ladder—Successes and Failures," *Research Management,* July 1977, pp. 30–33.
Connor, P. E., "Scientific Research Competence as a Function of Creative Ability," *IEEE Transactions on Engineering Management,* EM-21, 1, Feb 1974, pp. 2–8.
Edward, S. A. and McCarrey, M. W., "Measuring the Performance of Researchers," *Research Management,* Jan. 1973, pp. 34–41.
Epstein, K. A., "The Dual Ladder—Career Paths are not Always Equivalent," *IEEE Developing Careers: Issues for Engineers and Employers,* 1985.
Fischer, W. A. and McLaughlin, C. P., "MBO and R&D Productivity: Revisiting the System's Dynamics," *IEEE Transactions on Engineering Management,* EM-27, 4, Nov 1980, pp. 103–108.
Goldberg, A. I. and Shenhav, Y. A., "R&D Career Paths: Their Relation to Work Goals and Productivity," *IEEE Transactions on Engineering Management,* EM-31, 3, Aug 1984, pp. 111–117.
Jauch, L. R., "Tailoring Incentives for Researchers," *Research Management,* Nov. 1976, pp. 23–27.
Katz, R., ed., *Managing Professionals in Innovative Organizations,* Cambridge, MA: Ballinger, 1988.
Kleingartner, A. and Anderson, C., *Human Resource Management in High-Technology Firms,* Lexington, MA: Lexington Books, 1987.
Kopelman, R. E., "Merit Rewards, Motivation and Job Performance," *Research Management,* Jan. 1977, pp. 35–37.
Kornhauser, W., *Scientists in Industry,* Berkeley, CA: University of California Press, 1963.
Leonard-Barton, D., "The Case for Integrative Innovation: An Expert System at Digital," *Sloan Management Review,* 29, 1, Fall 1987, pp. 7–20.
Moore, D. C. and Davis, D. S., "The Dual Ladder—Establishing and Operating It," *Research Management,* July 1977, pp. 14–19.
Pelz, D. and Andrews, F., *Scientists in Organizations,* revised ed., University of Michigan Press, 1976.
Raftl, R. M., *R&D Productivity,* Culver City, CA: Hughes Aircraft Co. 1978.
Sacco, G. Jr. and Knopka, W., "Restructuring the Dual Ladder at Goodyear," *Research Management,* July–August 1983, pp. 36–41.
Schainblatt, A. H., "How Companies Measure the Productivity of Engineer and Scientists," *Research Management,* May 1982, pp. 10–18.
Schuster, J., *Management Compensation in High Technology Companies,* Lexington, MA: Lexington, 1984.

Shapero, A., "Managing Creative Professionals," *Research Management,* March–April 1985, pp. 23–28.

Shapira, R. and Globerson, S., "An Incentive Plan for R&D Workers," *Research Management,* Sept–Oct. 1983, pp. 17–20.

Shepard, H. A., "The Dual Hierarchy in Research," *Research Management,* pp. 177–187, Autumn 1958.

Shirley, E. and Michael, W. M., "Measuring the Performance of Researchers," *Research Management,* Jan. 1973, p. 34.

Smith, J. J. and Szabo, T. T., "The Dual Ladder—Importance of Flexibility, Job Content and Individual Temperament," *Research Management,* July 1977, pp. 20–23.

Stahl, M. J. and Steger, J. A., "Measuring Innovation and Productivity: A Peer Rating Approach," *Research Management,* Jan. 1977, pp. 35–38.

Stahl, M. J., Zimmerer, T. W. and Gulati, A., "Measuring Innovation, Productivity, and Job Performance of Professionals: A Decision Modeling Approach," *IEEE Transactions on Engineering Management,* EM-31, 1, Feb. 1984, pp. 25–29.

Wallmark, J. T. and Sedig, K. G., "Quality of Research Measured by Citation Method and by Peer Review—a Comparison," *IEEE Transactions on Engineering Management,* EM-33, 4, Nov. 1986, pp. 218–222.

Whitley, R. and Frost, P., "The Measurement of Performance in Research," *Human Relations,* 24, 2, 1971, pp. 161–177.

Wolff, M. F., "Misusing the Dual Ladder, or The Case of Joe Mertz," *Research Management,* March–April 1985, pp. 7–9.

Wolff, M. F., "Revisiting the Dual Ladder at General Mills," *Research Management,* May–June 1987, pp. 8–12.

2.1.4.2 Organization of technical resources [see also 3.1.3 and 3.1.5]

Allen, T. J. and Fusfeld, A. R., "Research Laboratory Architecture and the Structuring of Communications," *R&D Management,* 5, 2, 1975.

Allen, T. J., *Managing the Flow of Technology,* Cambridge, MA: MIT Press, 1977.

Allen, T. J., "Organizational Structure, Information Technology and R&D Productivity," *IEEE Transactions on Engineering Management,* EM-33, 4, 1986, pp. 212–217.

Badawy, M. K., "Organizational Designs for Scientists and Engineers: Some Research Findings and their Implications for Managers," *IEEE Transactions on Engineering Management,* EM-22, 4, Nov 1975, pp. 134–138.

Burns, T. and Stalker, G. M., *The Management of Innovation,* London: Tavistock, 1961.

Cohn, S. F. and Turyn, R. M., "The Structure of the Firm and the Adoption of Process Innovations," *IEEE Transactions on Engineering Management,* EM-27, 4, Nov 1980, pp. 98–102.

Ettlie, J. E., Bridges, W. P. and O'Keefe, R. D., "Organization Strategy and Structural Differences for Radical Versus Incremental Innovation," *Management Science,* 30, 6, June 1984: pp. 682–695.

Evans, G. S., "An Exploratory Test of the Matrix Assumption in a Highly Differentiated Research Organization: Structural Design Versus Behavioral Imperatives," *IEEE Transactions on Engineering Management,* EM-29, 3, Aug 1982, pp. 78–81.

Gerstenfeld, A., *Effective Management of Research and Development*, Reading, MA: Addison-Wesley, 1970.

Ginn, M. E., "Creativity Management: Systems and Contingencies from a Literature Review," *IEEE Transactions on Engineering Management*, EM-33, 2, May 1986, pp. 96–101.

Jerkovsky, W., "Functional Management in a Matrix Organizations," *IEEE Transactions on Engineering Management*, EM-30, 2, May 1983, pp. 89–97.

Johnson, S. and Jones, C., "How to Organize for New Products" *Harvard Business Review*, May–June 1957.

Kolodny, H., "Matrix Organizational Design and New Product Success," *Research Management*, Sept 1980, pp. 29–33.

Lawrence, P. R. and Lorsch, J. W., "New Management Job: Integrator," *Harvard Business Review*, Nov–Dec. 1967.

Leifer, R. and Triscari, T., Jr., "Research Versus Development: Differences and Similarities," *IEEE Transactions on Engineering Management*, EM-34, 2, May 1987, pp. 71–78.

Levinson, N. S. and Moran, D. D., "R&D Management and Organizational Coupling," *IEEE Transactions on Engineering Management*, EM-34, 1, Feb. 1987, pp. 28–35.

Lorsch, J. W., and Lawrence, D. R., "Organizing for Product Innovation," *Harvard Business Review*, January–February 1965, pp. 109–122.

Lorsch, J. W., *Product Innovation and Organization*, London: Collier-MacMillan, 1965, p. 184.

Morton, J. A., *Organizing for Innovation*, New York: McGraw-Hill, 1971.

Orth, C. D. and others, *Administering Research and Development: The Behavior of Scientists and Engineers in Organizations*, London, Tavistock, 1965.

Orth, C. D., 3rd, "The Optimum Climate for Industrial Research," *Harvard Business Review*, March–April 1959.

Perrow, C., *Normal Accidents*, New York: Basic, 1984.

Raab, M. D., "Meeting New Demands in Research Facilities," *Research Management*, March–April 1985.

Radosevich, R. and Robles, F., "Improving Matrix Relationships in R&D Organizations-Marketing Support Services," *R&D Management*, 14, 1984, pp. 229–232.

Roberts, E. B., "Stimulating Technological Innovation—Organization Approaches," *Research Management*, Nov 1979, pp. 26–30.

Rubenstein, A. H., "Organizational Factors Affecting Research and Development Decision-Making in Large Decentralized Companies," *Management Science*, Vol. 10, No. 4, July 1964, pp. 618–633.

Sullo, P. and Triscari, T., Jr. and Wallace, W. A., "Reliability of Communication Flow in R&D Organizations," *IEEE Transactions on Engineering Management*, EM-32-2, May 1985, pp. 91–97.

Szakonyi, R., "Keeping R&D Projects on Track," *Research Management*, Jan–Feb. 1985, pp. 29–34.

Thompson, J. D., *Organizations in Action*, New York: McGraw-Hill, 1967.

Tushman, M. L., "Managing Communication Networks in R&D Laboratories," *Sloan Management Review*, 20, 2, Winter 1979, pp. 27–49.

Walker, A. H. and Lorsch, J. W., "Organizational Choice: Product Versus Function," in

J. W. Lorsch and P. R. Lawrence, eds., *Studies in Organizational Behavior,* Ill: Irwin-Dorsey, 1970.

Wall, W. C., Jr., "Integrated Management in Matrix Organization," *IEEE Transactions on Engineering Management,* EM-31, 1, Feb 1984, pp. 30–36.

2.1.4.3 Funding

Bisio, A. and Gastwirt, L. E., "R&D Expenditures and Corporate Planning," *Research Management,* Jan. 1980, pp. 23–26.

Dohrmann, R. J., "Matching Company R&D Expenditures to Technological Needs," *Research Management,* Nov. 1978, pp. 17–21.

Dohrmann, R. J., "Quantifying Contract Research Projects," *Research Management,* 3, 1982.

Foster, R. N., *Innovation: The Attacker's Advantage,* New York: Summit, 1986.

Freeman, R. J., "A Stochastic Model for Determining the Size and Allocation of the Research Budget," *IEEE Transactions on Engineering Management,* EM-7, 1, 1960, pp. 2–7.

Graves, S. B., "Optimal R&D Expenditure Streams: An Empirical View," *IEEE Transactions on Engineering Management,* EM-34, 1, Feb. 1987, pp. 42–48.

Guerard, J. B. and Andrews, B. A., "R&D Management and Corporate Financial Policy," *Management Science,* 33, 11, 1987, pp. 1419–1429.

Guerard, J. B., Jr., Bean, A. S. and McCabe, G. M., "Identifying Intertemporal Relationships in Corporate R&D Expenditures," *IEEE Transactions on Engineering Management,* EM-33, 3, Aug. 1986, pp. 157–161.

Hanson, W. T., Jr. and Nason, H. K., "Funding and Budgeting R&D Equipment and Facilities," *Research Management,* Sept. 1980, pp. 25–28.

Kay, N. M., *The Innovating Firm: A Behavioral Theory of Corporate R&D,* London: MacMillan, 1979.

MacMillan, I. C., and Barbosa, B. R., "Business Unit Strategy and Changes in the Product R&D Budget," *Management Science,* 29, 7, 1983, pp. 757–769.

Madey, G. R. and Dean, B. V., "Strategic Planning for Investment in R&D Using Decision Analysis and Mathematical Programming," *IEEE Transactions on Engineering Management,* EM-32, 2, May 1985, pp. 84–90.

Mitchell, G. R. and Hamilton, W. F., "Managing R&D as a Strategic Option," *Research and Technology Management,* May–June 1988, pp. 15–22.

Naslund, B. and Sellstedt, B.,"An Evaluation of Some Methods for Determining the R&D Budget," *IEEE Transactions on Engineering Management,* EM-21, 1, Feb. 1974, pp. 24–29.

Radosevich, R. and Hayes, R. L., "Toward the Implementation of R&D Resource Allocation Models," *IEEE Transactions on Engineering Management,* EM-20, 1, Feb. 1973, pp. 32–33.

Research Management "How Much R&D?," June–July 1985, pp. 27–32.

Ronstadt, R., "A Method for Allocating R&D Expenditures," *IEEE Transactions on Engineering Management,* EM-12, Sept 1965, pp. 87–93.

2.1.4.4 Equipment in R,D&E

Adler, P. S. and Helleloid, D. A., "Effective Implementation of Integrated CAD/CAM: A Model," *IEEE Transactions on Engineering Management,* EM-34, 2, May 1987, pp. 101–107.

Adler, P. S., "Managing High-Tech Processes: The Challenge of CAD/CAM," in M. A. Von Glinow and S. Albers Mohrman, eds., *Managing Complexity in High Technology Industries, Systems and People,* New York: Oxford University Press, forthcoming.

Allen, T. J. and Hauptman, O., "The Influence of Communication Technologies on Organizational Structure," *Communication Research,* 14, 4, October 1987, pp. 575–587.

Beatty, C. A. and Gordon, J. R. M., "Barriers to the Implementation of CAD/CAM Systems," *Sloan Management Review,* Summer 1988.

Cooley, M., *Architect or Bee?* Boston: South End Press, 1980.

Fleischer, M., Liker, J. K. and Ansdorf, D. R., "Implementation and Use of Computer-Aided Design and Computer-Aided Engineering," *Transactions of 1987 IEEE Conference on Management of Evolving Systems,* Atlanta, Georgia.

Gagnon, R. J. and Mantel, S. J., Jr., "Strategies and Performance Improvement for Computer-Assisted Design," *IEEE Transactions on Engineering Management,* EM-34, 4, Nov 1987, pp. 223–235.

Hanss, E. J., "Are Decision Support Systems Applicable to Engineering Management?," *Engineering Management International,* 2, 1984, pp. 243–250.

Jaikumar, R. and Bohn, R. E., "The Development of Intelligent Systems for Industrial Use: An Empirical Investigation," in R. S. Rosenbloom, ed., *Research on Technological Innovation, Management and Policy,* 3, 1986.

Jones, P. E., "Technological Innovation in the Field of Automotive Diesel Engines," *IEEE Transactions on Engineering Management,* EM-23, 1, Feb. 1975, pp. 41–46.

Kargar, D. W. and Murdick, R. G., "A Management Information System for Engineering and Research," *IEEE Transactions on Engineering Management,* May 1977. 2148

Kiser, D. O. and Decker, C. D., "Improving R&D Effectiveness via Computers," *Research Management,* July–Aug. 1985, pp. 19–21.

Madey, G. R., Wolfe, M. H. and Potter, J., "Development of an Expert Investment Strategy System for Aerospace R,D&E and Production Contract Bidding," *IEEE Transactions on Engineering Management,* EM-34, 4, Nov. 1987, pp. 252–258.

Majchrzak, A., Chang, T. C., Barfied, W., Eberts, R., and Salvendy, G., *Human Aspects of Computer-Aided Design,* Philadelphia, PA: Taylor and Francis, 1987,

Manufacturing Studies Board, National Research Council, *Computer Integration of Engineering Design and Production: A National Opportunity,* Washington, DC: National Academy Press, 1984.

Myers, L. A. Jr., "Information Systems in Research and Development," *R&D Management,* 13, 4, 1983, pp. 199–206.

Schilling, P. E. and Lezark, A. P., "Planning for Laboratory Computing and Communications," *Research Management,* July–Aug. 1988, pp. 39–43.

Togna, A. D., "A Program for Support Operations," *Research Management,* Jan. 1977, pp. 30–34.

2.1.4.5 Interorganizational Linkages [see also 1.3.2, 1.5.4, 1.5.5]

2.1.4.5a Sourcing, commercializing, joint development and standards

Aiken, M. and Hage, J., "Organizational Interdependence and Intraorganizational Structure," *American Sociological Review,* 1968, pp. 912–930.

Anderson, B. A., "Marketing Techniques," *Research Management,* May 1979, pp. 26–28.

Baranson, J., "Technology Transfer Through the International Firm," *American Economic Review,* May 1970, pp. 435–440.

Barnes, C., "Get Invention Off the Shelf," *Harvard Business Review,* 44, 1, Jan–Feb. 1966.

Bell, J. R., "Patent Guidelines for Research Managers," *IEEE Transactions on Engineering Management,* EM-31, 3, Aug. 1984, pp. 102–104.

Berg, S. V., Duncan, J. and Friedman, P., *Joint Venture Strategies and Corporate Innovation,* Cambridge, MA: Oelgeschlager, Gunn & Hain, 1982.

Brown, J. H. U., "The Research Consortium—Its Organization and Functions," *Research Management,* May 1981, pp. 38–41.

Calvert, R., ed., *The Encyclopedia of Patent Practice and Invention Management,* New York: Reinhold; London: Chapman and Hall, 1964, p. 860.

Carpenter, M. P., Cooper, M. N. and Narin, F., "Linkage Between Basic Research Literature and Patents," *Research Management,* March 1980, pp. 30–35.

Ciborra, C., "*Strategic Alliance: Managing Corporations on the Fast Lane,*" Paper Presented at Symposium on "Managing Innovation in Large, Complex Firms," INSEAD, France, Aug. 31–Sept. 2, 1987.

Contractor, F., *International Technology Licensing: Compensation, Costs and Negotiations,* Lexington, MA: Lexington, 1981.

Davis, D. B., "R&D Consortia," *High Technology,* October 1985, pp. 42–52.

De Meuse, K. P. and Lounsbury, J. W., "Appraising the Performance of R&D Subcontractors," *Research Management,* Sept. 1981, pp. 32–37.

Dimancescu, D. and Botkin, J., *The New Alliance: America's R&D Consortia,* Cambridge, MA: Ballinger, 1986.

Dohrmann, R. J., "Quantifying Contract Research Projects," *Research Management,* 3, 1982.

Eaton, W., "Patent problem: Who owns the rights," *Harvard Business Review,* 45, 4, July–Aug. 1967.

Ford, D. and Ryan, C., "Taking Technology to Market," *Harvard Business Review,* March–April 1981.

Foster, R. N., "Organize for Technology Transfer," *Harvard Business Review,* Nov–Dec. 1971.

Friar, J. and Horwitch, M., "The Emergence of Technology Strategy: A New Dimension of Strategic Management," in M. Horwitch, ed., *Technology in Society,* 7, 2/3, 1986.

Friedman, P., Berg, S. V. and Duncan, J., "External vs Internal Knowledge Acquisition: Joint Venture Activity and R&D Intensity," *Journal of Economics and Business,* Winter 1979, pp. 103–108.

Fusfield, H. I., *The Technical Enterprise,* Cambridge, MA: Ballinger, 1986.

Fusfield, H. J. and Haklisch, C. S., "Cooperative R&D for Competitors," *Harvard Business Review,* November 1985, pp. 4–11.

Gabel, L., ed., *Product Standardization and Competitive Strategy,* Amsterdam: North Holland, 1987.

Gerstenfeld, A. and Berger, P., "Joint Research—A Wave of the Future?," *Research Management,* Nov–Dec. 1984, pp. 9–11.

Graham, M. B. W. and Rosenthal, SR., "*Institutional Aspects of Process Procurement for Flexible Machining Systems,*" Boston University, Manufacturing Roundtable Research Report, 1986.

Guetzkow, H., "Relations Among Organizations," in R. V. Bowers, ed., *Studies on Behavior in Organizations,* Athens, GA: University of Georgia Press, 1966, pp. 13–14.

Haklisch, C. A., Fusfield, H. J. and Levenson, A., "Trends in Collective Industrial Research," New York City Center for Technology Policy, December 1934.

Hamilton, W. F., "Corporate Strategies for Managing Emerging Technologies," in M. Horwitch, ed., *Technology in Society,* 7, 2/3, 1986.

Hariharan, S. and Prahalad, C. K., "*Technological Compatibility Choices in High-Tech Products: Implications for Corporate Strategy,*" Paper presented at the First Annual Conference on Managing the High-Tech Firm, Boulder, Colorado, Jan. 1988.

Hartley, F. L., "Technology Licensing for the Future," *Les Nouvelles,* March 1983, pp. 25–27.

Hayes, R. H. and Wheelwright, S. C., *Restoring Our Competitive Edge: Competing Through Manufacturing,* New York: John Wiley & Sons, 1984, Chapter 9.

Horwitch, M. and Prahalad, C. K., "Managing Multi-Organization Enterprises: The Emerging Strategic Frontier," *Sloan Management Review,* Winter 1981, pp. 3–16.

Horwitch, M., "*Changing Patterns for Corporate Strategy and Technology Management: The Rise of the Semiconductor and Biotechnology Industries.*" Paper Presented at the Mitsubishi Bank Foundation Conference on Business Strategy and Technical Innovations, March 1983, Itoh City, Japan.

Horwitch, M., "Designing and Managing Large-Scale, Public-Private Technological Enterprises: A State-of-the-Art Review," in M. Horwitch, *Technology in Society,* 1, 3, 1979, pp. 179–192.

Industrial Research Institute, "Position Statement on Licensing of Technology," *Research Management,* May 1979, pp. 32–33.

Jewkes, J., Sawers, D., and Stillerman, R., *The Sources of Invention,* New York, W. W. Norton & Co., 2nd ed, 1969.

Karlson, C., "*Corporate Families to Handle Galloping Technology,*" European Institute for Advanced Studies in Management, Working Paper 84-15, November 1984.

Killing, J. P., "Manufacturing Under License," *Business Quarterly,* Winter 1977.

Kitti, C. and Trozzo, C. L., *The Effects of Patent and Antitrust Laws, Regulations, and Practices on Innovation,* Virginia: Institute for Defense Analyses Program Analysis Division, 1976, 3, IDA Log No. HQ 76-18304.

Linn, R. A., "The When and Why of Licensing," *Research Management,* July, 1981, pp. 21–25.

Marcy, W., "Acquiring and Selling Technology-Licensing Do's and Don't," *Research Management,* May 1979, pp. 18–21.

McDonald, D. W. and Gieser, S. M., "Making Cooperative Research Relationships Work," *Research Management,* July–Aug. 1987, pp. 38–42.

McDonald, D. W. and Leahey, H. S., "Licensing has a Role in Technology Strategic Planning," *Research Management,* Jan–Feb. 1985, pp. 35–40.

Merrifield, B., "R&D Limited Partnerships are Starting to Bridge the Invention-Translation Gap," *Research Management,* May–June 1986, pp. 9–12.

Nemec, J. Jr., "Technology Commercialization: An Overview," in Booz, Allen & Hamilton: Outlook, Fall/Winter 1982.

Noyce, R. N., "Competition and Cooperation—A Prescription for the Eighties," *Research Management,* March 1982, pp. 13–17.

Ouchi, W. G. and Bolton, M. K., "The Logic of Joint Research and Development," *California Management Review,* 30, 3, Spring 1988, pp. 9–33.

Peck, M. J. and Goto, A., "Technology and Economic Growth: The Case of Japan," *Research Policy,* Nov 1981.

Peck, M. J., "Joint R&D: The Case of Microelectronics and Computer Technology Corporation," *Research Policy,* 15, 5, Oct. 1986, pp. 219–232.

Reddy, N. M., "Voluntary Product Standards: Linking Technical Criteria to Marketing Decisions," *IEEE Transactions on Engineering Management,* EM-34, 4, Nov. 1987, pp. 236–243.

Reich, L. S., "Research, Patents, and the Struggle to Control Radio: A study of Big Business and the Uses of Industrial Research," *Business History Review,* 1977.

Riggs, H. E., *Managing High-Technology Companies,* New York: Van Nostrand Reinhold, 1983.

Roberts, E. B. and Berry, C. A., "Entering New Businesses: Selecting Strategies for Success," *Sloan Management Review,* 26, 3, Spring 1985.

Roberts, E. B., "Is Licensing an Effective Alternative?," *Research Management,* Sept. 1982, pp. 20–24.

Ronstadt, R. and Kramer, R. J., "Getting the Most Out of Innovation Abroad," *Harvard Business Review,* March–April 1982.

Rosegger, G., "*Technological Incentives for Cooperation in the Auto Industry,*" Department of Economics, Case Western Reserve University, 1986.

Rothchild, R., "Making Patents Work for Small Companies," *Harvard Business Review,* 65, 4, Jul–Aug. 1987.

Senkins, M., "Licensing Sources and Resources," *Research Management,* May 1979, pp. 22–25.

Shanklin, W. L. and Ryans, J. K., Jr., *Marketing High Technology,* Lexington, MA: Lexington: Lexington, 1984.

Starr, C., "Industrial Cooperation in R&D," *Research Management,* September–October 1985, pp. 13–15.

Stinchcombe, A. L., "Contracts as hierarchical documents," in A. L. Stinchcombe and C. A. Heimer, *Organization Theory and Project Management,* Oslo: Norwegian University Press, 1985.

Teece, J. D., "Capturing Value from Technological Innovation: Integration, Strategic Partnering, and Licensing Decisions," *Research Policy,* 15, 1986, pp. 285–305.

Telesio, P., *Technology Licensing and Multinational Enterprises,* New York: Praeger, 1979.

Walker, G. and Weber, D., "A Transaction-Cost Approach to Make-or-Buy Decisions," *Administrative Science Quarterly,* 29, Sept 1984.

Walker, G., "Strategic Sourcing, Vertical Integration and Transaction Costs," *Interfaces,* 18, 3, May–June 1988, pp. 62–73.

Waugh, T. H., II, "Technology Centers Unite Industry and Academia," *High Technology,* October 1985.

Wells, L. T., Jr., *Third World Multinationals,* Cambridge, MA: MIT Press, 1983.

Zenoff, D. B., "Licensing as a Means of Penetrating Foreign Markets," *IDEA,* 14, Summer 1970.

2.1.4.5b External information [see also 3.1.3]

Allen, T. J., *Managing the Flow of Technology,* Cambridge, MA: MIT Press, 1977.

Baram, M. S., "Trade Secrets: What Price Loyalty," *Harvard Business Review,* Nov–Dec. 1968.

Brown, W. B. and Schwab, R. C., "Boundary-Spanning Activities in Electronics Firms," *IEEE Transactions on Engineering Management,* EM-31, 3, Aug. 1984, pp. 105–111.

Chakrabarti, A. K., Feineman, S. and Fuentevilla, W., "Characteristics of Sources, Channels, and Contents for Scientific and Technical Information Systems in Industrial R&D," *IEEE Transactions on Engineering Management,* EM-30, 2, May 1983, pp. 83–88.

Ciborra, C., "*Strategic Alliance: Managing Corporations on the Fast Lane,*" Paper Presented at Symposium on "Managing Innovation in Large, Complex Firms," INSEAD, France: Aug. 31–Sept. 2, 1987.

Ettlie, J. E., "The Timing and Sources of Information for the Adoption and Implementation of Production Innovations," *IEEE Transactions on Engineering Management,* EM-23, 1, Feb. 1975, pp. 62–68.

Fusfield, H. I. and Haklisch, C. S., *University Industry Research Interactions,* New York: Pergamon Press, 1984.

Ghoshal, S. and Kim, S. K., "Building Effective Intelligence Systems for Competitive Advantage," *Sloan Management Review,* Fall 1986, pp. 49–58.

Holland, W. E., Stead, B. A. and Leibrock, R. C., "Information Channel/Source Selection as a Correlate of Technical Uncertainty in a Research and Development Organization," *IEEE Transactions on Engineering Management,* EM-23, 4, Nov 1976, pp. 163–167.

Paolillo, J. G. P., "Technological Gatekeepers: A Managerial Perspective," *IEEE Transactions on Engineering Management,* EM-29, 4, Nov. 1982, pp. 169–171.

Rosenbloom, R. S. and Wolek, F. W., *Technology and Information Transfer,* Boston, MA: Harvard University Graduate School of Business Administration, 1970.

Von Hippel, E., "Cooperation Between Rivals: Informal Know-how Trading," *Research Policy,* 16, 1987, pp. 291–301.

Von Hippel, E., "Lead Users: A Source of Novel Product Concepts," *Management Science,* July 1986.

Von Hippel, E., "The Dominant Role of the User in Semiconductor and Electronic Subassembly Process Innovation," *IEEE Transactions on Engineering Management,* EM-24, 2, May 1977, pp. 60–71.

2.1.4.6 Project Selection and Termination

Baker, N. and Freeland, J., "Recent Advances in R&D Benefit Measurement and Project Selection Methods," *Management Science,* 21, 10, June 1985.

Baker, N. R. and Pound, W. H., "Project Selection: Where We Stand," *IEEE Transactions on Engineering Management,* EM-11, 4, Dec 1964.

Baker, N. R., "R&D Project Selection Models: An Assessment," *IEEE Transactions on Engineering Management,* EM-21, 4, Nov 1974, pp. 165–171.

Baker, R. N., Green, S. G. and Bean, A. S., "The Need for Strategic Balance in R&D Project Portfolios," *Research Management,* March–April 1986, pp. 38–43.

Balachandra, R. and Raelin, J. A., "How to Decide When to Abandon a Project," *Research Management,* July 1980, pp. 24–29.

Balachandra, R. "Go/NoGo Signals for New Product Development," *Journal of Product Innovation Management,* 1, 2, April 1984, pp. 92–100.

Balthasar, H. U., Boschi, R. A. A. and Menke, M. M., "Calling the Shots in R&D," *Harvard Business Review,* May–June 1978.

Bedell, R. J., "Terminating R&D Projects Prematurely," *Research Management,* July–Aug. 1983, pp. 32–35.

Buell, C. D., "When to Terminate a R&D Project," *Research Management,* July 1967, pp. 275–284.

Cooper, R. G., "An Empirically Derived New Product Project Selection Model," *IEEE Transactions on Engineering Management,* EM-28, 3, Aug. 1981, pp. 54–51.

Dean, J. W., Jr., *Deciding to Innovate: How Firms Justify Advanced Technology,* Cambridge, MA: Ballinger, 1987.

European Industrial Research Management Association, "Top-Down and Bottom-Up Approaches to Project Selection," *Research Management,* March 1978, pp. 22–24.

Gerwin, D., "Control and Evaluation in the Innovation Process: The Case of Flexible Manufacturing Systems," *IEEE Transactions on Engineering Management,* EM-28, 3, Aug. 1981, pp. 62–70.

Gold, B., "Strengthening Managerial Approaches to Improving Technological Capabilities," *Strategic Management Journal,* 4, 1983, pp. 209–222.

Hayes, R. H., Wheelwright, S. C. and Clark, K. B., *Dynamic Manufacturing,* New York: Wiley, 1988.

Hodder, J. E. and Riggs, H. E., "Pitfalls in Evaluating Risky Projects," *Harvard Business Review,* Jan–Feb. 1985, pp. 128–136.

Jaikumar, R., "Postindustrial Manufacturing," *Harvard Business Review,* Nov–Dec. 1986, pp. 69–76.

Kaplan, R. S., "Must CIM be Justified by Faith Alone?" *Harvard Business Review,* March–April 1986, pp. 87–95.

King, W. R., "The Strategic Evaluation of Projects and Programs," in B. V. Dean, ed., *Project Management,* Amsterdam: North-Holland, 1985.

Liberatore, M. J., "An Extension of the Analytic Hierarchy Process for Industrial R&D Project Selection and Resource Allocation," *IEEE Transactions on Engineering Management,* EM-34, 1, Feb 1987, pp. 12–18.

Meredith, J. R., ed., *Justifying New Manufacturing Technology,* Norcross GA: Industrial Engineering and Management Press, 1986.

Raelin, J. A. and Balachandra, R., "R&D Project Termination in High-Tech Industries," *IEEE Transactions on Engineering Management,* EM-32, 1, Feb 1985, pp. 16–23.

Souder, W. E., "Comparative Analysis of R&D Investment Models," *AIIE Transactions,* 4, 1972, pp. 57–64.
Souder, W. E., *Project Selection and Economic Appraisal,* New York: Van Nostrand, 1984.
Statman, M. and Tyebjee, T. T., "The Risk of Investment in Technological Innovation," *IEEE Transactions on Engineering Management,* EM-31, 4, Nov. 1984, pp. 165–171.
Steele, L. W., "Selecting R&D Programs and Objectives," *Research Technology Management,* March–April 1988, pp. 17–36.
White, G. R. and Graham, M. B. W., "How to Spot a Technological Winner," *Harvard Business Review,* March–April 1978.

2.1.4.7 Management of Technical Projects [see also 3.1.1,2,3,4,5, 3.2]

2.1.4.7a Development procedures

Abernathy, W. J. and Baloff, N., "Interfunctional Planning for New Product Introduction," *Sloan Management Review,* Winter 1972–73, pp. 25–43.
Abernathy, W. J. and Rosenbloom, R. S., "Parallel and Sequential R&D Strategies: Applications of a Simple Model," *IEEE Transactions on Engineering Management,* EM-15, 1, March 1968, pp. 2–10.
Abernathy, W. J., "Some Issues Concerning the Effectiveness of Parallel Strategies in R&D Projects," *IEEE Transactions on Engineering Management,* EM-18, 3, Aug 1971.
Badiru, A. B., *Project Management in Manufacturing and High Technology Operations,* New York: Wiley, 1988.
Batson, R. G., "Critical Path Acceleration and Simulation in Aircraft Technology Planning," *IEEE Transactions on Engineering Management,* EM-34, 4, Nov. 1987, pp. 244–251.
Bergen, S. A. and Pearson, A. W., "Project Management and Innovation in the Scientific Instrument Industry," *IEEE Transactions on Engineering Management,* EM-30, 4, Nov. 1983, pp. 194–199.
Booker, J. M. and Bryson, M. C., "Decision Analysis in Project Management: An Overview," *IEEE Transactions on Engineering Management,* EM-32, 1, Feb 1985, pp. 3–9.
Booz-Allen & Hamilton, "*Management of New Products,*" 1968.
Clark, K. B., Chew, B. W. and Fujimoto, T., "Product Development in the World of Auto Industry: Strategy, Organization and Performance," *Brookings Papers on Economic Activity,* 3, 1987, pp. 729–771.
Cochran, E. B., Patz, A. L., and Rowe, A. J., "Concurrency and Disruption in New Product Innovation," *California Management Review,* 21, 1, Fall 1978, pp. 21–33.
Cooper, R. G. and Kleinschmidt, E. J., "An Investigation into the New Product Process: Steps, Deficiencies, Impact," *Journal of Product Innovation Management,* 3, 1986, pp. 72–85.
Cooper, R. G., "A Process Model for Industrial New Product Development," *IEEE Transactions on Engineering Management,* EM-30, 1, Feb 1983, pp. 2–11.

Crosby, P. B., *Quality is Free,* New York: Mentor, 1980.

Department of Defense, Assistant Secretary of Defense, Acquisition and Logistics, *Transition from Development to Production,* DOD 4245.7-M, Sept 1985.

Eastman, R. M., ''Engineering Information Release Prior to Final Design Freeze,'' *IEEE Transactions on Engineering Management,* EM-27, 2, May 1980, pp. 37–42.

Gardiner, P. and Rothwell, R., *Innovation: A Study of the Problems and Benefits of Product Innovation,* London: Design Council, 1985.

Gerstenfeld, A., *Effective Management of Research and Development,* Reading, MA: Addison-Wesley, 1970.

Gluck, F. W. and Foster, R. N., ''Managing Technological Change: A Box of Cigars for Brad,'' *Harvard Business Review,* Sept–Oct. 1975.

Gryna, F. M., ''Product Development,'' in J. M. Juran and F. M. Gryna, eds., *Quality Control Handbook,* 4th ed., New York: McGraw-Hill, 1988.

Hayes, R. H., Wheelwright, S. C. and Clark, K. B., *Dynamic Manufacturing,* New York: Wiley, 1988.

Horwitch, M., ''Managing Large-Scale Programs: The Managerial Dilemma,'' *Technology in Society,* 6, 1984.

Imai, K., Nonaka, I. and Takeuchi, H., ''Managing the New Product Development Process: How Japanese Learn and Unlearn,'' in K. B. Clark, R. H. Hayes and C. Lorenz, eds., *The Uneasy Alliance: Managing the Productivity-Technology Dilema,* Boston, MA: Harvard Business School Press, 1985.

Johnson, S. C. and Jones, C., ''How to Organize for New Products,'' *Harvard Business Review,* May–June 1957, pp. 49–62.

Kozar, K. A., ''Team Product Reviews: A Means of Improving Product Quality and Acceptance,'' *Journal of Product Innovation Management,* 4, 1987, pp. 204–216.

Kraft, C. L. II, ''Concurrency: Schedule Compression is Quality's Challenge,'' *Quality Progress,* Sept 1983, pp. 12–17.

Lee-Kwang, H. and Farrel, J., ''The SSD Graph: A Tool for Project Scheduling and Visualization,'' *IEEE Transactions on Engineering Management,* EM-35, 1, Feb. 1988, pp. 25–30.

Liker, J. K. and Hancock, W. M., ''Organizational Barriers to Engineering Effectiveness,'' *IEEE Transactions on Engineering Management,* EM-33, 2, May 1986, pp. 82–91.

Love, S. F., *Planning and Creating Successful Engineered Designs: Managing the Design Process,* North Hollywood, CA: Advanced Professional Development, 1986.

Mansfield, E., Rapoport, J., Schnee, J., Wagner, S. and Hamburger, M., *Research and Innovation in the Modern Corporation,* New York: Norton, 1971.

McDonough, E. F. III and Spital, F. C., ''Quick Response New Product Development,'' *Harvard Business Review,* Sept–Oct. 1984.

Might, R., ''An Evaluation of the Effectiveness of Project Control Systems,'' *IEEE Transaction on Engineering Management,* EM-31, 3, Aug. 1984, pp. 127–137.

Moder, J. J., Phillips, C. R. and Davis, E. W., *Project Management with CAM, PERT, and Procedence Diagramming,* New York: Van Nostrand, 1983.

Reinertsen, D. G., ''Blitzkrieg Product Development: Cut Development Times in Half,'' *Electronic Business,* Jan. 15, 1985.

Rosennau, M. D. Jr., ''From Experience: Faster New Product Development,'' *Journal of Production Innovation Management,* 5, 1988, pp. 150–153.

Scherer, F. M., "Time-Cost Tradeoffs in Uncertain Empirical Research Projects," *Naval Research Logistics Quarterly,* 13, March 1966; pp. 71–82.
Schmidt-Tiedemann, K. J., "A New Model of the Innovation Process," *Research Management,* March 1982, pp. 18–21.
Souder, W., *Managing New Product Innovations,* Lexington, MA: Lexington, 1987.
Stalk, G. Jr., "Time—The Next Source of Competitive Advantage," *Harvard Business Review,* July–Aug 1988, pp. 41–53.
Takeuchi, H. and Nonaka, T., "The New New Product Development Game," *Harvard Business Review,* Jan–Feb. 1986.
Whelan, J. M., "Project Profile Reports Measure R&D Effectiveness," *Research Management,* Sept. 1976, pp. 14–16.

2.1.4.7b Implementation procedures [see also 3.1.5]

Adler, P. S., "Automation and Skill: New Directions," *International Journal of Technology Management,* 2, 5–6, 1987, pp. 761–772.
Adler, P. S., "New Technologies, New Skills," *California Management Review,* 29, 1, Fall 1986, pp. 9–28.
Argote, L., Goodman, P. S. and Schkade, D., "The Human Side of Robotics: How Workers React to a Robot," *Sloan Management Review,* 24, 1983, pp. 31–41.
Boddy, D. and Buchanan, D. A., *Managing New Technology,* Oxford: Basil Blackwell, 1986.
Braverman, H., *Labor and Monopoly Capital,* New York: Monthly Review Press, 1974.
Bright, J. R., "Does Automation Raise Skill Requirements?" *Harvard Business Review,* 36, July–Aug. 1958, pp. 85–98.
Buchanan, D. A. and Boddy, D., *Organizations in the Computer Age: Technological Imperatives and Strategic Choice,* Aldershot: Gower, 1983.
Cornfield, D. B., ed., *Workers, Managers and Technological Change,* New York: Plenum, 1987.
de Pietro, R. A. and Schremser, G. M., "The Introduction of Advanced Manufacturing Technology (AMT) and its Impact on Skilled Workers' Perception of Communication, Interaction, and other Job Outcomes at a Large Manufacturing Plant," *IEEE Transactions on Engineering Management,* EM-34, 1, Feb 1987, pp. 4–11.
Gerwin, D., "Do's and Don'ts of Computerized Manufacturing," *Harvard Business Review,* March–April 1982, pp. 107–116.
Hawthorne, E. P., *The Management of Technology,* London: McGraw-Hill 1978.
Hayes, R. H., Wheelwright, S. C., Clark and K. B., *Dynamic Manufacturing,* New York: Wiley, 1988.
Hirschhorn, L., *Beyond Mechanization,* Cambridge, MA: MIT Press, 1984.
Howard, R., *Brave New Workplace,* New York: Viking, 1985.
Imada, A. and Noro, K., eds., *Participatory Ergonomics,* Philadelphia, PA: Taylor and Francis, forthcoming.
Jaikumar, R. and Bohn, R. E., "The Development of Intelligent Systems for Industrial Use: An Empirical Investigation," in R. S. Rosenbloom, ed., *Research on Technological Innovation, Management and Policy,* 3, 1986.
Jaikumar, R., "Post-Industrial Manufacturing," *Harvard Business Review,* Nov–Dec. 1986, pp. 69–76.

Kennedy, D., Craypo, C. and Lehman, M., eds., *Labor and Technology: Union Responses to Changing Environments,* Dept. of Labor Studies, Pennsylvania State University, 1982.

Knights, D., Willnott, H. and Collinson, D., eds., *Job Redesign: Critical Perspectives on the Labour Process,* Hampshire, UK: Gower, 1985.

Kujawa, D., "Technology Strategy and Industrial Relations: Case Studies of Japanese Multinationals in the United States," *Journal of International Business Studies,* Winter 1983.

Lansbury, R. D. and Davis, E. M., *Technology, Work and Industrial Relations,* Melbourne, Australia: Longman Cheshire, 1984.

Leonard-Barton, D., "*Implementation as Mutual Adaption of Technology and Organization,*" Harvard Business School Working Paper, 88-016, 1987.

Leydesdorff, L. and Zeldenrust, S., "Technological Change and Trade Unions," *Research Policy,* 13, 3, June 1984.

Mumford, E. et al., "The Human Problems of Computer Introduction," *Management Decision,* Vol. 10, Spring 1972, pp. 6–17.

Noble, D. F., "Social Choice in Machine Design: The Case of Automatically Controlled Machine Tools," in A. Zimbalist, ed., *Case Studies on the Labor Process,* New York: Monthly Review Press, 1979.

Pasmore, W. A. and Sherwood, J. J. eds., *Sociotechnical Systems: A Sourcebook,* LaJolla, CA: University Associates, 1978.

Pava, C., *Managing New Office Technology,* New York: Free Press, 1983.

Perrow, C., "The Organizational Context of Human Factors Engineering," *Administrative Science Quarterly,* 28, 4, 1983, pp. 521–541.

Sandberg, A., ed., *Computers Dividing Man and Work: Recent Scandinavian Research on Planning and Computers from a Trade Union Perspective,* Stockholm: Arbetslivscentrum, 1979.

Slichter, S. H., Healy, J. J. and Livernash, E. R., *The Impact of Collective Bargaining on Management,* Washington, DC.: Brookings, 1960.

Walton, R. E. and McKersie, R. B., "*Managing New Technology and Labor Relations: An Opportunity for Mutual Influence,*" MIT Management in the 1990s Working Paper 88-062, Oct 1988.

Walton, R. E. and Susman, G. I., "People Policies for New Machines," *Harvard Business Review,* March–April 1987.

Walton, R. E., "Challenges in the Management of Technology and Labor Relations," in R. E. Walton and P. R. Lawrence, eds., *HRM Trends and Challenges,* Boston, MA: Harvard Business School Press, 1985.

Walton, R. E., "From Control to Commitment: Transforming Workforce Management in the United States," in K. B. Clark, R. H. Hayes and C. Lorenz eds., *The Uneasy Alliance: Managing the Productivity-Technology Dilemma,* Boston; MA: Harvard Business School Press 1985.

Wood, S., ed., *The Degradation of Work? Skill, Deskilling and the Labour Process,* London: Hutchinson, 1982.

Zuboff, S., *In the Age of the Smart Machine,* New York: Basic, 1988.

Zuboff, S., "Technologies that Informate: Implications for Human Resource Management in the Computerized Industrial Workplace," in R. E. Walton and P. R. Lawrence, eds., *HRM Trends and Challenges,* Boston MA: Harvard Business School Press, 1985.

2.1.4.7c Staffing technical projects

Brooks, F. P., Jr., *The Mythical Man-Month,* Reading, MA: Addison-Wesley, 1982.

Dill, D. D. and Pearson, A. W., "The Effectiveness of Project Managers: Implications of a Political Model," *IEEE Transactions on Engineering Management,* EM-31, 3, Aug. 1984, pp. 138–146.

Frohman, A. L., "Mismatch Problems in Managing Professionals," *Research Management,* Sept. 1978, pp. 20–25.

Helms, C. P. and Wyskida, R. M., "A Study of Temporary Task Teams," *IEEE Transactions on Engineering Management,* EM-31, 2, May 1984, pp. 55–60.

Katz, R. and Allen, T. J., "Investigating the Not Invented Here (NIH) Syndrome: A Look at the Performance, Tenure and Communication Patterns of 50 R&D Project Groups," *R&D Management,* 12, 1, 1982.

Katz, R., "As Research Teams Grow Older," *Research management,* Jan–Feb. 1984, pp. 23–28.

Katz, R., "The Effects of Group Longevity on Project Communication and Performance," *Administrative Science Quarterly,* 27, 1982, pp. 81–104.

Kidder, T., *The Soul of A New Machine,* Boston: Little, Brown, 1981.

Kruytbosch, C. E., "Management Styles and Social Structure in 'Identical' Engineering Groups," *IEEE Transactions on Engineering Management,* ME-9, 3, Aug. 1972, pp. 92–102.

Might, R. J., "The Role of Structural Factors in Determining Project Management Success," *IEEE Transactions on Engineering Management,* EM-32, 2, May 1985, pp. 71–77.

Moore, W. L. and Tushman, M. L., "Managing Innovation over the Product Life Cycle," in M. L. Tushman and W. L. Moore eds., *Readings in the Management of Innovation,* Marshfield, MA: Pitman, 1982.

Pearson, A. W., "A Model for Studying Some Organizational Effects of An Increase in the Size of R&D Projects," *IEEE Transactions on Engineering Management,* EM-26, 1, Feb. 1979, pp. 14–21.

Posner, B. Z., "What's All the Fighting About? Conflicts in Project Management," *IEEE Transactions on Engineering Management,* EM-33, 4, Nov. 1986, pp. 207–211.

Smith, C. G., "Scientific Performance and the Composition of Research Teams," *Administrative Science Quarterly,* 16, 1971, pp. 486–495.

Thamhaim, H. J. and Wilemon, D. L., "Building High Performing Engineering Project Teams," *IEEE Transactions on Engineering Management,* EM-34, 3, Aug 1987, pp. 130–137.

Wallmark, J. T. and Sellerberg, B., "Efficiency vs. Size of Research Teams," *IEEE Transactions on Engineering Management,* EM-13, 3, Sept. 1966, pp. 137–142.

Wallmark, J. T., Eckerstein, S., Langered, B. and Holmqvist, H. E. S., "The Increase in Efficiency with Size of Research Teams," *IEEE Transactions on Engineering Management,* EM-20, Aug. 1973, pp. 80–86.

Woleck, F. W., "Engineering Roles in Development Projects," *IEEE Transactions on Engineering Management,* EM-19, 2, May 1972, pp. 53–60.

2.1.4.8 *Managing Ongoing Technological Operations*

2.1.4.8a *Quality control in technical operations*

Burgess, J. A., "Design Assurance—A Tool for Excellence," *Engineering Management International,* 5, 1, April 1988, pp. 25–30.

Burgess, J. A., *Design Assurance for Engineers and Managers,* New York: Marcel Decker, 1984.

De Cotiis, T. A. and Dyer, L., "Defining and Measuring Project Performance," *Research Management,* Jan. 1979, pp. 17–22.

Hollocker, C. P., "Finding the Cost of Software Quality," *IEEE Transactions on Engineering Management,* EM-33, 4, Nov. 1986, pp. 223–228.

Kraft, C. L. II, "Concurrency: Schedule Compression is Quality's Challenge," *Quality Progress,* Sept. 1983, pp. 12–17.

McDonough, E. F. and Kinnunen, R. M., "Management Control of New Product Development Projects," *IEEE Transactions on Engineering Management,* EM-31, Feb. 1984, pp. 18–21.

Melan, E. H., "Quality Improvement in an Engineering Laboratory," *Quality Progress,* June 1987, pp. 18–25.

Murray, T. J., "Meeting the New Quality Challenge," *Research Management,* Nov–Dec. 1987, pp. 25–30.

Nichols, A. E., "Measuring Research: Quality in Creating," *ASQC Quality Congress Transactions,* Dallas, 1988.

Perrow, C., *Normal Accidents,* New York: Basic, 1984.

Rymer, J. E., "Managing the Process-The Easy Way to Improve Profits," *IEEE Transactions on Engineering Management,* EM-31, 2, May 1984, pp. 92–95.

Schrader, L. J., "The Engineering Organization's Cost of Quality Program," *Quality Progress,* Jan. 1986, pp. 29–34.

Takei, F., "Engineering Quality Improvement Through TQC Activity," *IEEE Transactions on Engineering Management,* EM-33, 2, May 1986, pp. 92–95.

Tribus, M., "Applying Quality Management Principles," *Research Management,* Nov–Dec. 1987, pp. 11–21.

Wolff, M. F., "Quality/Process Control: What R&D Can Do," *Research Management,* Jan–Feb. 1976, pp. 9–11.

2.1.4.8b *Schedule and cost control of technical operations [see also 2.1.4.8d]*

Bower, J. L. and Hout, T. M. "Fast-Cycle Capability for Competitive Power," *Har Business Review,* Nov–Dec 1988, pp. 110–118.

Liker, J. K. and Hancock, W. M., "Organizational Barriers to Engineering Effec ness," *IEEE Transactions on Engineering Management,* EM-33, 2, May 1986, 82–91.

Robles, F. and Radosevich, R., "Needs Assessment for Support Units in an Organization," *IEEE Transactions on Engineering Management,* EM-31, 2, 1984, pp. 70–75.

Stanton, G. B., Jr., "How to Control the Use of Staff Time," *Research Management,* Sept. 1977, pp. 21–24.

2.1.4.8c Regulatory compliance [see also 1.3.4]

Deane, R. H., "Minimizing Legal Liability for Unsafe Products," *IEEE Transactions on Engineering Management,* EM-22, 4, Nov. 1975, pp. 144–149.

Hauptman, O. and Roberts, E. B., "FDA Regulation of Product Risk and its Impact Upon Young Biomedical Firms," *Journal of Product Innovation Management,* 4, 2, June 1987.

Ochs, D. L., "Complying with Human and Animal Research Regulations," *Research Management,* 9, 1981.

Stanton, G. B., Jr., "Health and Safety Challenges: The Engineering Manager's Response," *IEEE Transactions on Engineering Management,* EM-28, 3, Aug. 1981, pp. 76–78.

2.1.4.8d R,D&E performance measurement

Andrews, F. M., ed., *Scientific Productivity,* London: Cambridge University Press and Unesco, 1979.

Brown, M. G. and Svenson, R. A., "Measuring R&D Productivity," *Research Management,* July–Aug. 1988, pp. 11–15.

Collier, D. W. and Gee, R. E., "A Simple Approach to Post-Evaluation of Research," *Research Management,* May 1973, pp. 12–17.

Collier, D. W., "Measuring the Performance of R&D Departments," *Research Management,* May 1977, pp. 30–34.

De Meuse, K. P. and Lounsbury, J. W., "Appraising the Performance of R&D Subcontractors," *Research Management,* Sept. 1981, pp. 32–37.

Dukes, R. E., Dyckman, T. R. and Elliott, J. A., "Accounting for Research and Development Costs: The Impact on Research and Development Expenditures," *Journal of Accounting Research,* Vol. 18, Supplement, 180, p. 1–37.

Fusfield, H. I. and Langlois, R. N., *Understanding R&D Productivity,* New York: Pergamon Press, 1981.

Gambino, A. J. and Gartenberg, M., "Costing and Reporting R&D Operations," *Research Management,* July 1979, pp. 15–18.

Hodge, M. H., "Rate Your Company's Research Productivity," *Harvard Business Review,* Nov–Dec. 1963, pp. 109–122.

Lipetz, B., *The Measurement of Efficiency of Scientific Research,* Carlisle, MA: Intermedia, 1965.

Mechlin, G. F. and Berg, D., "Evaluating Research—ROI is not Enough," *Harvard Business Review,* Sept–Oct. 1980.

Narin, F., Carpenter, M. P. and Woolf, P., "Technological Performance Assessments Based on Patents and Patent Citations," *IEEE Transactions on Engineering Management,* EM-31, 4, Nov. 1984, pp. 172–183.

Packer, M. B., "Analyzing Productivity in R&D Organizations," *Research Management,* Jan–Feb. 1983, pp. 13–20.

Pappas, R. A. and Remer, D. S., "Measuring R&D Productivity," *Research Management,* May–June 1985, pp. 15–22.

Patterson, W. C., "Evaluating R&D Performance at Alcoa Laboratories," *Research Management,* March–April 1983, pp. 23–27.

Quinn, J. B., "How to Evaluate Research Output," *Harvard Business Review,* March-April 1960.

Raftl, R. M., *R&D Productivity,* Culver City, CA: Hughes Aircraft Co. 1978.

Richard, A. P. and Donald S. R., "Measuring R&D Productivity," *Research Management,* May–June 1985.

Rubenstein, A. H., "Setting Criteria for R&D," *Harvard Business Review,* 35, 1, 1957, pp. 95–104.

Takei, F., "Evaluation Method for Engineering Activity—One Example in Japan," *IEEE Transactions on Engineering Management,* EM-28, 1, Feb. 1981, pp. 13–16.

Takei, F., "Evaluation Method for Engineering Activity Through Comparison with Competition—Four Years' Experience," *IEEE Transactions on Engineering Management,* EM-32, 2, May 1985, pp. 63–70.

Vinkler, P., "Management System for a Scientific Research Institute Based on the Assessment of Scientific Publications," *Research Policy,* 15, 2, April 1986, pp. 77–88.

2.1.4.9 Interfunctional Interfaces

2.1.4.9a With manufacturing [see also 3.1.5]

Abernathy, W. J. and Wayne, K., "Limits of the Learning Curve," *Harvard Business Review,* September–October 1974.

Abernathy, W. J., *The Productivity Dilemma: Roadblock to Innovation in the Automobile Industry,* Baltimore: Johns Hopkins University Press, 1978.

Bergen, S. A., "Analytical Instrument R&D Management," *R&D Management,* 2, 1, 1971.

Bergen, S. A., *Productivity and the R&D/Production Interface,* Hampshire, Gower, 1983.

Burgess, J. A., *Design Assurance for Engineers and Managers,* New York: Marcel Decker, 1984.

Department of Defense, Assistant Secretary of Defense, Acquisition and Logistics, *Transition from Development to Production,* DOD 4245.7-M, Sept. 1985.

Dwivedi, S. N. and Klein, B. R., "Design for Manufacturability Makes Dollars and Sense," *CIM Review,* 2, 3, Spring 1986, pp. 53–59.

Etienne, E. C., "Interactions Between R&D and Process Technology," *Research Management,* Jan. 1981, pp. 22–27.

Ettlie, J. E. and Reifeis, S. A., "Integrating Design and Manufacturing to Deploy Advanced Manufacturing Technology," *Interfaces,* 17, 6, Dec. 1987, pp. 63–74.

Francis, P., "A Review of Selected Weapons' Transition to Production," U.S. General Accounting Office, Sept 1984.

Gerstenfeld, A., *Effective Management of Research and Development,* Reading, MA: Addison-Wesley, 1970.

Gold, B., "Alternate Strategies for Advancing a Company's Technology," *Research Management*, July 1975.

Hayes, R. H. and Wheelwright, S. C., *Restoring Our Competitive Edge: Competing through Manufacturing*, New York: John Wiley & Sons, 1984, Chapter 10.

Hayes, R. H., Wheelwright, S. C., Clark and K. B., *Dynamic Manufacturing*, New York: Wiley, 1988.

Hutchinson, J., "Evolving Organizational Forms," *Columbia Journal of World Business*, 11, 2, Summer 1976.

International Management, "When Engineers Talk to Each Other—the Slow but Sure Payoff," July 1984.

Jones, J. C., "Continuous Design and Redesign," *Design Studies*, Jan. 1983, pp. 53–60.

McCann, J. F. and Galbraith, J. R., "Interdepartmental Relations," in P. P. Nystrom and W. H. Starbuck, eds., *Handbook of Organizational Design*, Vol. 1, New York: Oxford University Press, 1981, pp. 60–84.

Production, "Manufacturing/Design Teamwork—Producibility is Key to Profitability," May 1979.

Quinn, J. B. and Mueller, J. A., "Transferring Research Results to Operations," *Harvard Business Review*, Jan–Feb. 1963.

Riggs, H. E., *Managing High-Technology Companies*, New York: Van Nostrand Reinhold, 1983.

Roussel, P. A., "Cutting Down the Guesswork in R&D," *Harvard Business Review*, Sept–Oct. 1983, pp. 154–160.

Rubenstein, A. H. and Ginn, M. E., "Project Management at Significant Interfaces in the R&D/Innovation Process," in B. V. Dean, ed., *Project Management*, New York: North-Holland, 1985.

Scheirer, M., "Approaches to the Study of Implementation," *IEEE Transactions on Engineering Management*, EM-30, 2, May 1983, pp. 76–82.

Sherpa, J. R., "Manufacturing and Design Engineering—Partnership in Cost Reduction," *IEEE Transactions on Manufacturing Technology*, Dec. 1972, pp. 13–14.

Szakonyi, R., "Don't Drop the Ball—Or Your New Product," *Research and Development*, Nov 1987, pp. 66–68.

Szakonyi, R., "Worlds Apart—Bridging R&D and Manufacturing," *Manufacturing Engineering*, Dec. 1987, pp. 67–70.

White, J. A., *Design for Automation—A System View*, Georgia Institute of Technology, 1983.

Wolff, M. F., "Bridging the R&D Interface with Manufacturing," *Research Management*, Jan–Feb. 1985, pp. 9–11.

2.1.4.9b With marketing

Berenson, C., "The R&D Marketing Interface: A General Analogue Model for Technology Diffusion," *Journal of Marketing*, 32, 1968, pp. 8–15.

Bonnet, D. C.-L., "Nature of the R&D/Marketing Cooperation in the Design of Technologically Advanced New Industrial Products," *R&D Management*, 16, 2, 1986, pp. 117–126.

Dougherty, D. J., "Thought Worlds and Organizing as an Unnatural Act: An Innovation-

Focussed Look at Differentiation and Integration,'' Department of Management, Wharton School, University of Pennsylvania, WP 807, 1988.

Gupta, A. K., Raj, S. P. and Wilemon, D. L., ''R&D and Marketing Dialogue in High-Tech Firms,'' *Industrial Marketing Management,* 14, 1985, pp. 289–300.

Gupta, A. K., Raj, S. P. and Wilemon, D. L., ''R&D and Marketing Managers in High-Tech Companies: Are They Different?'' *IEEE Transactions on Engineering Management,* EM-33, 1, Feb. 1986, pp. 25–32.

Link, A. N. and Zmud, R. W., ''Additional Evidence on the R&D/Marketing Interface,'' *IEEE Transactions on Engineering Management,* EM-33, 1, Feb. 1986, pp. 43–43.

Montelone, J. P., ''How R&D and Marketing Can Work Together,'' *Research Management,* Nov. 1975, pp. 19–21.

Myers, S. and Marquis, D. G., ''Successful Innovation,'' *National Science Foundation,* NSF, 67-17, 1969.

Radosevich, R. and Robles, F., ''Improving Matrix Relationships in R&D Organizations—Marketing Support Services,'' *R&D Management,* 14, 1984, pp. 229–232.

Riggs, H. E., *Managing High-Technology Companies;* New York: Van Nostrand Reinhold, 1983.

Rothwell, R., Freeman, C., Horsley, A., Jervis, V. T. P., Robertson, A. B. and Townsend, J., ''SAPPHO Updated—Project SAPPHO, Phase II,'' *Research Policy,* 3, 1974, pp. 258–291.

Shanklin, W. L. and Ryans, Jr., J. K., ''Organizing for High-Tech Marketing,'' *Harvard Business Review,* Nov–Dec. 1984, pp. 164–171.

Souder, W. E., and Chakrabarti, A. K., ''The R&D/Marketing Interface: Results from an Empirical Study of Innovation Projects,'' *IEEE Transactions on Engineering Management,* EM-25, 4, Nov. 1978, pp. 88–93.

Souder, W. E., ''Effectiveness of Nominal and Interacting Group Decision Process for Integrating R&D and Marketing,'' *Management Science,* Feb. 1977, pp. 595–605.

Souder, W. E., ''Effectiveness of Product Development Methods,'' *Industrial Marketing Management,* 7, 1978.

Souder, W. E., ''Promoting Effective R&D/Marketing Interfaces,'' *Research Management,* July 1980, pp. 10–15.

Sullivan, L. P., ''Quality Function Deployment,'' *Quality Progress,* June 1986, pp. 39–50.

Szakonyi, R., ''Dealing with a Nonobvious Source of Problems Related to Selecting R&D to Meet Customers' Future Needs: Weaknesses Within an R&D Organization's and Within a Marketing Organization's Individual Operations,'' *IEEE Transactions on Engineering Management,* EM-35, 1, Feb 1988, pp. 37–41.

Von Hippel, E., ''Lead Users: A Source of Novel Product Concepts,'' *Management Science,* July 1986.

Von Hippel, E., ''The Dominant Role of Users in the Scientific Instrument Innovation Process,'' *Research Policy,* 5, 3, 1976, pp. 212–239.

Weinrauch, J. D. and Anderson, R., ''Conflicts Between Engineering and Marketing Units,'' *Industrial Marketing Management,* 11, 1982, pp. 291–301.

Young, H. C., ''Effective Management of Research-Market Teams,'' *Research Management,* March 1979, pp. 7–12.

2.1.4.10 Inter-divisional Relations

2.1.4.10a Sharing resources [see also 2.3.2]

Bartlett, C. A. and Ghoshal, S., "Managing Across Borders: New Strategic Requirements," *Sloan Management Review,* Summer 1987, pp. 7–17.

Doz, Y., Angelmar, R. and Prahalad, C. K., "Technological Innovation and Interdependence: A Challenge for the Large Complex Firm," in M. Horwitch, ed., *Technology in Society,* 7, 2/3, 1986, pp. 14–34.

Ghoshal, S. and Bartlett, C. A., "*Creation, Adoption and Diffusion of Innovations by Subsidiaries of Multinational Corporations,*" Paper Presented at Symposium on Managing Innovation in Large, Complex Firms, INSEAD, France, Aug. 31–Sept. 2, 1987.

Ghoshal, S. and Bartlett, C. A., "*Innovation Processes in Multinational Corporations,*" Paper Presented at Symposium on Managing Innovation in Large, Complex Firms, INSEAD, France, Aug. 31–Sept. 2, 1987.

Gupta, A. K. and Govandaranjan, V. "Resource Sharing Among SBUs," *Academy of Management Journal,* 29, 4, 1986, pp. 695–714.

Harrigan, K. R., "Managing Innovation Within Overseas Subsidiaries," *Journal of Strategy,* Spring 1984.

Hedlund, G., "The Modern MNC—A Heterarchy?" *Human Resource Management,* 25, 1, 1986, pp. 9–35.

Lorsch, J. W., Allen, S. A., III., *Managing Diversity and Interdependence,* Boston, MA: Division of Research, Harvard Graduate School of Business Administration, Harvard University, 1973.

Mansfield, E. and Romero, A., "'Reverse' Transfers of Technology from Overseas Subsidiaries to American Firms," *IEEE Transactions on Engineering Management,* EM-31, 3, Aug. 1984, pp. 122–127.

Pitts, R. A., "Strategies and Structures for Diversification," *Academy of Management Journal,* 20, 2, 1977, pp. 197–208.

Porter, M. E., *Competitive Advantage,* New York: Free Press, 1985.

Salter, M. S., "Tailor Incentive Compensation to Strategy," *Harvard Business Review,* Mar–April 1973, pp. 94–102.

Vancil, R. F., *Decentralization: Managerial Ambiguity by Design,* New York: Financial Executives Research Foundation, 1980.

Westwood, A. R. C., "R&D Linkages in a Multi-Industry Corporation," *Research Management,* May–June 1984, pp. 23–26.

2.1.4.10b Formation of new lines of business [see also 2.2.2]

Burgelman, R. A. and Sayles, L. R., *Inside Corporate Innovation,* New York: Free Press, 1986.

Burgelman, R. A., "A Process Model of Internal Corporate Venturing in a Diversified Major Firms," *Administrative Science Quarterly,* 28, 1983, pp. 223–244.

Burgelman, R. A., "Corporate Entrepreneurship and Strategic Management: Insights from a Process Study," *Management Science,* 29, 1983, pp. 1349–1364.

Burgelman, R. A., "Managing Corporate Entrepreneurship: New Structures for Implementing Technological Innovation," in M. Horwitch, ed., *Technology in Society*, 7, 2/3, 1986, pp. 91–104.

Burgelman, R. A., "Managing the Internal Corporate Venturing Process: Some Recommendations for Practice," *Sloan Management Review*, Winter 1984, pp. 33–48.

Burgelman, R. A., "Managing the New Venture Division: Research Findings and Implications for Strategic Management," *Strategic Management Journal*, 6, 1985, pp. 39–54.

Fast, N. D., "A Visit to the New Venture Graveyard," *Research Management*, March 1979.

Fast, N. D., "The Future of Industrial New Venture Departments," *Industrial Marketing Management*, 8, 1979, pp. 264–273.

Fast, N. D., *The Rise and Fall of Corporate New Venture Divisions*, Ann Arbor, MI: UMI Research Press, 1979.

Gee, E. A. and Tyler, C., *Managing Innovation*, New York: John Wiley and Sons, 1976.

Martin, M. J. C., *Managing Technological Innovation and Entrepreneurship*, Reston, VT: Reston, 1984.

Roberts, E. B., "New Ventures for Corporate Growth," *Harvard Business Review*, July/August 1980, pp. 134–142.

Von Hippel, E., "Successful and Failing Internal Corporate Ventures: An Empirical Analysis," *Industrial Marketing Management*, 6, 1977, pp. 163–174.

2.1.4.10c Relations between corporate R&D and divisions [see also 3.1.5]

Cohen, H., Keller, S. and Streeter, D., "The Transfer of Technology from Research to Development," *Research Management*, May 1979, pp. 11–17.

Ounjian, M. L. and Carne, E. B., "A Study of the Factors which Affect Technology Transfer in a Multilocation Multibusiness unit Corporation," *IEEE Transactions on Engineering Management*, EM-34, 3, Aug. 1987, pp. 194–201.

Roberts, E. B. and Frohman, A., "Strategies for Improving Research Utilization," *Technology Review*, 80, 5, March–April, 1978.

Smith, J. J., McKeon, J. E., Hoy, K. L., Boysen, R. L., Shechter, L. and Roberts, E. B., "Lessons from 10 Case Studies in Innovation," *Research Management*, 27, 5, Sept–Oct. 1984.

Stewart, J. M., "Techniques for Technology Transfer Within the Business Firm," *IEEE Transactions on Engineering Management*, EM-16, 3, Aug. 1969, pp. 103–110.

White, W., "Effective Transfer of Technology from Research to Development," *Research Management*, Jan. 1977, pp. 30–34.

2.1.5 Technology Strategy and Functional Strategies

Adler, P. S. and Ferdows, K., "*The Role of the Chief Technical Officer*," Stanford University, Industrial Engineering and Engineering Management Working Paper, Dec 1988.

Bourgeois, J. L., III. and Browdin, D. R., "Strategic Implementation: Five Approaches to an Elusive Phenomenon," *Strategic Management Journal*, 5, 1984, pp. 241–264.

Davis, D. D., "Integrating Technological, Manufacturing, Marketing and Human Resource Strategies," in D. D. Davis and Assoc. *Managing Technological Innovations,* San Francisco: Jossey-Bass, 1986.

Guth, W. D., MacMillan, J. C., "Strategy Implementation vs Middle-Management Self-Interest," *Strategic Management Journal,* 7, July 1986.

Hayes, R. H. and Wheelwright, S. C., *Restoring Our Competitive Edge: Competing Through Manufacturing,* New York: John Wiley & Sons, 1984.

Hayes, R. H., Wheelwright, S. C. and Clark, K. B., *Dynamic Manufacturing,* New York: Wiley, 1988.

Newman, W. H., Logan, J. P. and Hegarty, W. H., *"Multi-Level Strategy: Fusing Departmental Programs with Strategic Moves,"* Working Paper 57, Strategy Research Center, Graduate School of Business, Columbia University, 1987.

Willyard, C. H. and McClees, C., "Motorola's Technology Roadmap Process," *Research Management,* Sept–Oct. 1987, pp. 13–19.

2.1.5.1 R,D&E Functional Strategy [see also 2.1.1]

Bean, A. S., Baker, N. R., Green, S. G., Blank, W. and Srinivasan, V., "Involvement of R&D in Corporate Strategic Planning: Effects on Selected R&D Management Practices," in B. Bozeman, M. Crow and A. Link, eds., *Strategic Management of Industrial R&D,* Lexington, MA: D. C. Heath, 1984.

Cathey, P., "Technology's Input is Vital to Sound Business Planning," *Iron Age,* September 16, 1981, pp. 43–48.

Chambers, J. C., Emerald, R. L. and Rubenstein, A., "Coupling Corporate Strategy and R&D Planning," *Managerial Planning,* May–June 1985, pp. 35–48.

Crawford, C. M., "Defining the Charter for Product Innovation," *Sloan Management Review,* Fall 1980, pp. 3–12.

Cutler, G. W., "Formulating the Annual Research Program at Whirlpool," *Research Management,* January 1979, pp. 23–26.

Dean, B. V. and Goldhar, J. L., eds., *Management of Research and Innovation,* Oxford: North-Holland, 1980.

Drucker, P. F., "Twelve Fables of Research Management," *Harvard Business Review,* Jan–Feb. 1963.

Eschenback, T. G. and Geistauts, G. A., "Strategically Focused Engineering: Design and Management," *IEEE Transactions on Engineering Management,* EM-34, 2, May 1987, pp. 62–70.

Finkin, E. F., "Developing and Managing New Products," *Journal of Business Strategy,* 3, 1983, pp. 39–46.

Foster, R. N., "Boosting the Payoff from R&D," *Research Management,* January 1982.

Foster, R. N., *Innovation: The Attacker's Advantage,* New York: Summit, 1986.

Foster, R. N., "Linking R&D to Strategy," *The McKinsey Quarterly,* Winter 1981.

Frohman, A. L. and Bitondo, D., "Coordinating Business Strategy and Technical Planning," *Long Range Planning,* 14, Dec. 1981, pp. 58–67.

Frohman, A. L., "Technology as a Competitive Weapon," *Harvard Business Review,* Jan–Feb. 1982.

Fusfield, A. R., "How to Put Technology into Corporate Planning," *Technology Review,* May 1978.

Gerstenfeld, A., *Effective Management of Research and Development,* Reading, MA: Addison-Wesley, 1970.

Graham, M. B. W., "Corporate Research and Development: The Latest Transformation," in M. Horwitch ed., *Technology in Society,* 7, 2/3, 1986, pp. 86–101.

Graham, M. B. W., "Industrial Research in the Age of Big Science," in R. S. Rosenbloom, ed., Research on Technological Innovation, *Management and Policy,* 2, 1985, pp. 47–49.

Graham, M. B. W., "Radical Innovation: Ends versus Means at Alcoa and RCA," Boston University, 1984.

Graham, M. B. W., *The Business of Research: RCA and the Videodisc,* New York: Cambridge University Press, forthcoming.

Hambrick, D. C. and MacMillan, I. C., "Efficiency of Product R&D in Business Units: The Role of Strategic Context," *Academy of Management Journal,* 28, 3, 1985, pp. 527–547.

Hampbel, R. G., "Building the Total Corporate R&D Effort at Alcoa," *Research Management,* January 1979, pp. 27–30.

Hanson, W. T., Jr., "Planning R&D at Eastman Kodak," *Research Management,* July 1978, pp. 23–25.

Hettinger, W. D., "The Top Technologist Should Join the Team—The Corporate Officer and Board Team, That Is," *Research Management,* March 1982, pp. 7–10.

Hull, F. M., Hage, J., and Azumi, K., "R&D Management Strategies: America Versus Japan," *IEEE Transactions on Engineering Management,* EM-32, 2, May 1985, pp. 78–83.

Kay, N. M. and Freeman, C., *The Innovating Firm: A Behavioral Theory of Corporate R&D,* London: MacMillan, 1979.

Liberatore, M. J. and Titus, G. J., "Synthesizing R&D Planning and Business Strategy: Some Preliminary Findings," *R&D Management* 13, 1983, pp. 207–218.

Link, A. N. and Tassey, G., *Strategies for Technology-Based Competition,* Lexington, MA: Lexington, 1987.

Linn, R. A., "A Sectoral Approach to Strategic Planning for R&D," *Research Management,* January–February 1983.

MacAvoy, T. C., "The R&D Factor in Corporate Strategy," in *Corporate R&D Strategy, Innovation and Funding Issues,* Conference Board Report no. 863, New York: The Conference Board, Inc., 1985.

McGlauchlin, L., "Long-Range Technical Planning," *Harvard Business Review,* July–August 1968, pp. 54–64.

Miller, T. R., "Planning R&D at Union Carbide," *Research Management,* January 1978, pp. 31–33.

Morone, J. and Alben, R., "Matching R&D to Business Needs," *Research Management,* Sept–Oct. 1984, pp. 33–39.

Morton, J. A., *Organizing for Innovation,* New York: McGraw-Hill, 1971.

Mowery, D. C., "The Relationship Between Intrafirm and Contractual Forms of Industrial Research in America Manufacturing," *Explorations in Economic History,* 1983b.

Petroni, G., "Strategic Planning and Research and Development—Can We Integrate Them?,," *Long Range Planning,* 16, 1983, pp. 15–25.

Petroni, G., "Who Should Plan Technological Innovation?" *Long Range Planning,* 18, 1985, pp. 108–115.

Pruitt, B. H. and Smith, G. D., "The Corporate Management of Innovation: Alcoa Research, Aircraft Alloys and the Problem of Stress Corrosion Cracking," in R. S. Rosenbloom, ed., *Research on Technological Innovation, Management and Policy,* 3, 1987.

Quinn, J. B., "Fundamental Research Can be Planned," *Harvard Business Review,* Jan–Feb. 1964.

Quinn, J. B., "Long-Range Planning of Industrial Research," *Harvard Business Review,* July–Aug. 1961.

Rabino, S., and Moskowitz, H., "The R&D Role in Bringing New Products into the Marketplace," *Journal of Business Strategy,* 1, 1981, pp. 26–32.

Reich, L., *The Making of American Industrial Research: Science and Business at G.E. and Bell, 1876–1926,* New York: Cambridge University Press, 1986.

Roberts, E. B., *The Dynamics of Research and Development,* New York, Harper, 1964.

Rosenbloom, R. S. and Kantrow, A. M., "The Nurturing of Corporate Research," *Harvard Business Review,* Jan–Feb. 1982. 2151

Rosenbloom, R. S., "*The R&D Pioneers, Then and Now,*" Paper presented at the Hagley Museum and Library Conference on "The R&D Pioneers," Delaware, October 1985.

Rubinger, B., "Technology Policy in Japanese Firms: Decision-Making, Supplier Links and Technical Goals," in M. Horwitch ed., *Technology in Society,* 7, 2/3, 1986, pp. 184–200.

Rubenstein, A. H. and Geisler, E. "Long Range Planning of Corporate Research and Development" *Planning Review,* March 1985, pp. 39–41.

Ruggles, R. C., "How to Integrate R&D and Corporate Goals," *Management Review,* September 1982, pp. 8–17.

Schmitt, R. W., "Successful Corporate R&D," *Harvard Business Review,* May–June 1985.

Scholz, C., "Planning Procedures in German Companies—Finding & Consequences," *Long Range Planning,* 17, 1984, pp. 94–103.

Szakonyi, R., "To Improve R&D Productivity, Gain the CEO's Support," *Research Management,* May–June 1985, pp. 6–7.

2.1.5.2 Technology and Manufacturing Strategy

Abernathy, W. J., *The Productivity Dilemma: Roadblock to Innovation in the Automobile Industry,* Baltimore: Johns Hopkins University Press, 1978.

Buffa, E. S., *Meeting the Competitive Challenge: Manufacturing Strategy for U.S. Companies,* Homewood, IL: Dow Jones-Irwin, 1984.

Ferdows, K., Miller, J. G., Nakane, J. and Vollman, T. E., "Evolving Manufacturing Strategies in Europe, Japan and North America," Boston, MA.: Boston University, 1985.

Ferdows, K., "Technology Push Strategies for Manufacturing," *Tijdschrift voor Economie en Management,* 27, 2, 1983.

Flaherty, M. T., "Coordinating International Manufacturing and Technology," in M. E. Porter, ed., *Competition in Global Industries,* Boston, MA: Harvard Business School Press, 1986.

Gold, B., "Alternate Strategies for Advancing a Company's Technology," *Research Management,* July 1975.

Gold, B., "CAM Sets New Rules for Production," *Harvard Business Review*, Nov–Dec 1982, pp. 88–94.

Gunn, T. G., *Manufacturing for Competitive Advantage*, Cambridge, MA: Ballinger 1987.

Hayes, R. H., Wheelwright, S. C. and Clark, K. B., *Dynamic Manufacturing*, New York: Wiley, 1988.

Hayes, R. H. and Wheelwright, S. C., *Restoring Our Competitive Edge: Competing through Manufacturing*, New York: John Wiley & Sons, 1984.

Jelinek, M. and Goldhar, J. D., "The Interface Between Strategy and Manufacturing Technology," *Columbia Journal of World Business*, Spring 1983.

Kleindorfer, P. R., ed., *The Management of Productivity and Technology in Manufacturing*, New York: Plenum, 1986.

Meredith, J. R., "Strategic Planning for Factory Automation by the Championing Process," *IEEE Transactions on Engineering Management*, EM-33, 4, Nov. 1986, pp. 229–232.

Rafii, F., "Upgrading the Manufacturing Function's Strategic Role," *Operations Management Review*, Fall 1984.

Roth, A. and Miller, J. G., "Manufacturing Futures Fact Book: 1987 North American Manufacturing Futures Survey," Manufacturing Roundtable, Boston University, 1987.

Skinner, W., *Manufacturing: The Formidable Competitive Weapon*, New York: Wiley, 1985.

Stobaugh, R. and Telesio, P., "Match Manufacturing Policies and Product Strategy," *Harvard Business Review*, March–April, 1983.

Thompson, H. and Paris, M., "The Changing Face of Manufacturing Technology," *The Journal of Business Strategy*, 3, 1, Summer 1982.

2.1.5.3 *Information Systems Strategy*

Benjamin, R. I., Rockart, J. F., Scott-Morton, M. S., and Wyman, J., "Information Technology: A Strategic Opportunity," *Sloan Management Review*, 25, 3, Spring 1984, pp. 3–10.

Business Week, "Management's Newest Star: Meet the Chief Information Officer," October 13, 1986.

Camillus, J. C. and Lederer, A. L., "Corporate Strategy and Design of Computerized Information Systems," *Sloan Management Review*, Spring 1985, pp. 35–42.

Cash, J. I., Jr. and Konsynski, B. R., "IS Redraws Competitive Boundaries," *Harvard Business Review*, March–April 1985, pp. 134–142.

Cash, J. I., MacFarlan, F. W. and McKenney, J. L., *Corporate Information System Management: The Issues Facing Senior Executives*, 2nd edition, Homewood, Illinois: Dow Jones-Irwin, 1988.

Child, J., "Information Technology, Organization and the Response to Strategic Challenges," *California Management Review*, Fall 1987, pp. 33–50.

Gerstein, M. S., *The Technology Connection: Strategy and Change in the Information Age*, Reading, MA: Addison-Wesley, 1987.

Gibson, C. F. and Jackson, B. B., *The Information Imperative*, Lexington, MA: Lexington, 1987.

Gremillion, L. L. and Pyburn, P., "Breaking the Systems Development Bottleneck," *Harvard Business Review,* March–April 1983, pp. 130–137.
King, W. R., "Strategic Planning for MIS," *MIS Quarterly,* 2, 1, March 1978, pp. 27–37.
Lucas, H. C., Jr. and Turner, J. A., "A Corporate Strategy for the Control of Information Processing," *Sloan Management Review,* Spring 1982, pp. 25–36.
McFarlan, F. W., ed., *The Information Systems Research Challenge,* Boston, MA: Harvard Business School Press, 1984.
McFarlan, F. W. and McKenney, J. L., *Corporate Information Systems Management,* Homewood, IL: Irwin, 1982.
McFarlan, F. W., "Information Technology Changes the Way You Compete," *Harvard Business Review,* May–June 1984, pp. 98–103.
McFarlan, F. W., McKenney, J. L. and Pyburn, P., "Information Archipelago—Charting the Course," *Harvard Business Review,* Jan–Feb. 1983.
McKenney, J. L. and McFarlan, F. W., "The Information Archipelago—Maps and Bridges," *Harvard Business Review,* Sept–Oct. 1982, pp. 109–119.
Nolan, R. L., "Managing Information Systems by Committee," *Harvard Business Review,* July–Aug. 1982, pp. 72–79.
Office Systems, July 1987, pp. 273–289.
Parsons, G. L., "Information Technology: A New Competitive Weapon," *Sloan Management Review,* 25, 1, Fall 1983, pp. 3–14.
Rackoff, N., Wiseman, C., and Ulrich, W. A., "Information Systems for Competitive Advantage: Implementation of a Planning Process," *MIS Quarterly,* 9, 4, December 1985, pp. 285–294.
Rockart, J. F., "The Changing Role of the Information Systems Executive: A Critical Success Factors Perspective," *Sloan Management Review,* Fall 1982, pp. 2–13.
Sullivan, C. H., Jr. "Systems Panning in the Information Age," *Sloan Management Review,* Winter 1985, pp. 3–12.
Wyman, J., "Technological Myopia—The Need to Think Strategically About Technology," *Sloan Management Review,* Summer 1985, pp. 59–64.

2.1.5.4 *Technology and Marketing Strategy*

Abell, D., *Competitive Market Strategies,* Cambridge, MA: Marketing Science Institute, 1975.
Abell, D., *Strategic Market Planning,* Englewood Cliffs, NJ: Prentice-Hall, 1979.
Abrikian, H. S., "New Product Innovation: Empirical Findings and Marketing Applications," Stanford University, Graduate School of Business, Technical Report 67, June 1981.
Brockhoff, K. and Chakrabarti, A. K., "R&D/Marketing Linkages and Innovation Strategy: Some West Germany Experiences," Drexel University College of Business and Administration, Working Paper 88-1, 1988.
Day, G. S., *Strategic Market Planning,* St. Paul, MN: West, 1984.
Divita, S. F., "Selling R&D to the Government," *Harvard Business Review,* Sept–Oct. 1965.
Kiel, G., "Technology and Marketing: The Magic Mix?," *Business Horizons,* May–June 1974.

Kotler, P., Fahey, L., and Jatusripitak, S., *The New Competition,* Englewood Cliffs, NJ: Prentice-Hall, 1985.
Levitt, T., "Exploit the Product Life Cycle," *Harvard Business Review,* Nov–Dec. 1965.
McKenna, R., "Market Positioning in High Technology," *California Management Review,* Spring 1985.
Midgley, D., *Innovation and New Product Marketing,* New York: Halsted, 1977.
O'Shaughnessy, J. *Competitive Marketing: A Strategic Approach,* Boston: George Allen and Unwin, 1984.
Porter, M. E., *Competitive Advantage,* New York: Free Press, 1985.
Shanklin, W. L. and Ryans, J. K., Jr., *Marketing High Technology,* Lexington, MA: Lexington: Lexington, 1984.
Smilor, R. W., ed., *Customer-Driven Marketing: Lessons from Entrepreneurial Technology Companies,* Lexington, MA: Lexington, D.C. Heath, 1989.
Urban, G. L. and Hauser, J. R., *The Design and Marketing of New Products,* Englewood Cliffs, N.J.: Prentice-Hall, 1980.
Wind, J., Grasahof, J. and Goldhar, J., "Market-Based Guidelines for Design of Industrial Products," *Journal of Marketing,* 42, 1978, pp. 27–37.

2.1.5.5 Technology and Human Resource Strategy

Butler, J. E., Ferris, G. R., Schellenberg Smith Cook, D., "Exploring Some Critical Dimensions of Strategic Human Resource Management," in R. S. Schuler and S. A. Young, eds., *Readings in Personnel and Human Resource Management,* St. Paul, MN: West, 1987.
Dyer, L., "Bringing Human Resources into the Strategy Formulation Process," *Human Resource Management,* 22, 1983, pp. 257–271.
Fombrun, C. J., Tichy, N. M. and Devanna, M. A., eds., *Strategic Human Resource Management,* New York: John Wiley & Sons, 1984.
Schuler, R. and Jackson, S., "Linking Competitive Strategies with Human Resource Management Practices," *Academy of Management Executive,* 1, 3, 1987, pp. 207–219.
Walton, R. E. and McKersie, R. B., "Managing New Technology and Labor Relations: An Opportunity for Mutual Influence," MIT Management in the 1990s Working Paper 88-062, Oct 1988.

2.2 Technology Strategy Implementation and Organizational Structure

2.2.1 Theoretical Models of Environment-Organization Linkages

Barley, S. R., "Technology as an Occasion for Structuring: Evidence from Observations of CT Scanners and the Social Order of Radiology Departments," *Administrative Science Quarterly,* 31, 1986, pp. 78–108.
Barley, S. R., "Technology, Power and the Social Organization of Work: Towards a Pragmatic Theory of Skilling and Deskilling," in S. Bacharach and N. Ditomaso, eds., *Research in the Sociology of Organizations,* 6, 1988.

Benson, K., "The Interorganizational Network as a Political Economy," *Administrative Science Quarterly,* 20, 1975, pp. 229–249.

Burns, T. and Stalker, G. M., *The Management of Innovation,* London: Tavistock, 1961.

Child, J., "Organizational Structure, Environment, and Performance: The Role of Strategic Choice," *Sociology,* 6, 1972, pp. 2–22.

Cyert, R. M. and March, J. G., *A Behavioral Theory of the Firm,* Englewood Cliffs, NJ: Prentice-Hall, 1963.

Dill, W. R., "Environment as an Influence on Managerial Autonomy," *Administrative Science Quarterly,* 2, 1958, pp. 409–443.

Gerwin, D., "Relationship Between Structure and Technology" in P. C. Nystrom and W. H. Starbuck, eds., *Handbook of Organizational Design,* London: Oxford University Press, 1981.

Hage, J. and Aiken, M., "Routine Technology, Social Structure and Organizational Goals," *Administrative Science Quarterly,* 14, 1969, pp. 366–376.

Hickson, D. J., Pugh, D. S., and Phesey, D., "Operations Technology and Organization Structure: an Empirical Reappraisal," *Administrative Science Quarterly,* 14, 1969, pp. 378–397.

Hirsch, P. M., "Organizational Effectiveness and the Institutional Environment," *Administrative Science Quarterly,* 20, September 1975, pp. 327–344.

Hitt, M. A., Ireland, R. D. and Patia, K. A., "Industrial Firms' Grand Strategy and Importance: Moderating Effects of Technology and Uncertainty," *Academy of Management Journal,* 25, 1982, pp. 265–298.

Lawrence, P. R. and Lorsch, J. W., *Organization and Environment,* Boston, MA: Graduate School of Business Administration, Harvard University, 1967.

March, J. G. and Olsen, J. P., *Ambiguity and Choice in Organizations,* Bergen, Norway: Universitetsforlaget, 1976.

Meyer, J. W. and Rowan, B., "Institutionalized Organizations: Formal Structure as Myth and Ceremony," *American Journal of Sociology,* 83, 1977, pp. 340–363.

Miles, R. E., Snow, C. C. and Pfeffer, J., "Organization-Environment: Concepts and Issues," *Industrial Relations,* 13, 1974, pp. 244–264.

Miller, A., "Technology, Strategy, Performance; What Are the Links?" in B. Bozeman, M. Crow and A. Link, eds., *Strategic Management of Industrial R&D,* Lexington, MA: D. C. Health, 1984.

Ouchi, W. G., "Markets, Bureaucracies and Clans," *Administrative Science Quarterly,* 25, 1, 1980, pp. 129–141.

Perrow, C., "A Framework for Comparative Organizational Analysis," *American Sociological Review,* 32, 1967, pp. 194–208.

Pfeffer, J. and Leblebici, H., "The Effects of Competition on Some Dimensions of Organizational Structure," *Social Forces,* 52, 1973, pp. 268–279.

Pfeffer, J. and Nowak, P., "Joint Ventures and Interorganizational Dependence," *Administrative Science Quarterly,* 21, 1976, pp. 398–418.

Pfeffer, J. and Salanick, G. R., *The External Control of Organizations: A Resource Dependence Perspective,* New York: Harper & Row, 1978.

Pfeffer, J., *Organizations and Organization Theory,* Boston, MA: Pitman, 1982.

Scott, W. R., *Organizations: Rational, Natural and Open Systems,* Englewood Cliffs, NJ: Prentice-Hall, 1986.

Scott, W. R., "*Technology and Organizations: An Organizational Level Perspective,*"

Paper Presented at the Conference on Technology and Organizations, Graduate School of Industrial Administration, Carnegie Mellon University, Aug 28–30, 1988

Thompson, J. D. and Bates, J. F., "Technology, Organization, and Administration," *Administrative Science Quarterly,* 1967, pp. 325–342.

Thompson, J. D. and McEwen, W. J., "Organizational Goals and Environment," *American Sociological Review,* 23, 1958, pp. 23–31.

Thompson, J. D., *Organizations in Action,* New York, McGraw-Hill, 1967.

Thompson, V. A., "Bureaucracy and Innovation," *Administrative Science Quarterly* 10, 1965, pp. 1–20.

Van de Ven, A. H. and Drazin, R., "The Concept of Fit in Contingency Theory," in B. M. Staw and L. L. Cummins, eds., *Research in Organizational Behavior,* 7, 1985, pp. 333–365.

Van de Ven, V. H., Angle, H. and Poole, M. S., *Research on the Management of Innovation,* Cambridge, MA: Ballinger, 1988.

Weick, K. E., "Cognitive Processes in Organizations" in B. M. Staw, ed., *Research in Organizational Behavior,* 1, 1979, pp. 41–74.

Woodward, J., *Industrial Organization: Theory and Practice,* London: Oxford University Press, 1965.

2.2.2 Structures For Innovation [see also 2.1.4.2]

Allen, J. W., "Technology Resource Management," *Booz, Allen & Hamilton: Outlook,* Fall/Winter 1981.

Allen, T. J., *Managing the Flow of Technology,* Cambridge, MA: MIT Press, 1977.

Burgelman, R. A., "Managing Corporate Entrepreneurship: New Structures for Implementing Technological Innovation," in M. Horwitch, ed., *Technology in Society,* 7, 2/3, 1986, pp. 91–104.

Burns, T. and Stalker, G. M., *The Management of Innovation,* London: Tavistock, 1961.

Chandler, A. D., Jr., *Strategy and Structure,* Cambridge, MA: MIT Press, 1962.

Dougherty, D. J., "*Thought Worlds and Organizing as an Unnatural Act: An Innovation-Focussed Look at Differentiation and Integration,*" Department of Management, Wharton School, University of Pennsylvania, WP 807, 1988.

Ettlie, J. E., Bridges, W. P. and O'Keefe, R. D., "Organization Strategy and Structural Differences for Radical Versus Incremental Innovation," *Management Science,* 30, 6, June 1984: pp. 682–695.

Flaherty, M. T., "Coordinating International Manufacturing and Technology," in M. E. Porter, ed., *Competition in Global Industries,* Boston, MA: Harvard Business School Press, 1986.

Galbraith, J. R., *Designing Complex Organizations,* Reading, MA: Addison-Wesley, 1973.

Goldman, J. E., "Innovation in Large Firms," in R. S. Rosenbloom, ed., *Research on Technological Innovation, Management and Policy,* 2, 1985, pp. 1–10.

Horwitch, M. and Thietart, R. A., "The Effect of Business Interdependencies on Product R&D-Intensive Business Performance," *Management Science,* 33, 2, Feb. 1987, pp. 178–197.

Jelinek, M., *Institutionalizing Innovation,* New York: Praeger, 1979.

Kanter, R. M., *The Change Masters: Innovation for Productivity in the American Corporation,* New York: Simon & Shuster, 1983.

Lorsch, J. W., and Lawrence, P. R., "Organizing for Product Innovation," *Harvard Business Review,* January–February 1965, pp. 109–122.

Ouchi, W. G., "Markets, Bureaucracies and Clans," *Administrative Science Quarterly,* 25, 1, 1980.

Pascale, R. and Athos, A. *The Art of Japanese Management,* New York: Warner Books, 1981.

Penrose, E. T., *The Theory of the Growth of the Firm,* White Plains, NY: M. E. Sharpe, 1980.

Peters, T. J. and Waterman, R. H. Jr., *In Search of Excellence,* New York: Harper and Row, 1982.

Peterson, R. A., "Entrepreneurship and Organization" in P. Nystom and W. Starbuck, eds., *Handbook of Organizational Design,* New York: Oxford University Press, 1981.

Quinn, J. B., "Managing Innovation: Controlling Chaos," *Harvard Business Review,* May–June 1985.

Roberts, E. B., "Facts and Folklore in Research and Development Management," *Industrial Management Review,* 8, 2, 1967, pp. 5–18.

Roberts, E. B., "Generating Effective Corporate Innovation," *Technology Review,* October–November 1977, pp. 25–33.

Roberts, E. B., "Stimulating Technological Innovation—Organizational Approaches," *Research Management,* November 1979, pp. 26–30.

Roberts, E. B., "Toward A New Theory for Research and Development," *Industrial Management Review,* 4, 1, Fall 1962, pp. 29–40.

Rosenbloom, R. S. and Cusumano, M. A., "Technological Pioneering: The Birth of the VCR Industry," *California Management Review,* Summer 1987, pp. 51–76.

Rubenstein, A. H., and Ginn, M. E., "Project Management at Significant Interfaces in the R&D/Innovation Process," in B. V. Dean, ed., *Project Management,* New York: North-Holland, 1985.

Tushman, M. L., "Communications Across Organizational Boundaries: Special Boundary Roles in the Innovation Process," *Administrative Science Quarterly,* 22, 1977.

Von Hippel, E., "The Dominant Role of the User in Semiconductor and Electronic Subassembly Process Innovation," *IEEE Transactions on Engineering Management,* EM-24, 2, May 1977, pp. 60–71.

Von Hippel, E., "The Dominant Role of Users in the Scientific Instrument Innovation Process," *Research Policy,* 5, 3, 1976, pp. 212–239.

Von Hippel, E. A., "Successful Industrial Products from Customer Ideas," *Journal of Marketing,* January 1978.

2.3 Technology Strategy Process

Frederickson, J. W., "Strategic Process Research: Questions and Recommendations," *Academy of Management Review,* 8, 4, 1983, pp. 565–575.

Hax, A. C. and Majluf, N. S., "Defining Strategy and the Strategy Formation Process," *Interfaces,* May–June 1988, pp. 99–109.

Huff, A. S. and Reger, R. K., "A Review of Strategy Process Research," *Journal of Management,* 13, 2, 1987, pp. 211–236.
Quinn, J. B., "Innovation and Corporate Strategy," in M. Horwitch, ed., *Technology in Society,* 7, 2/3, 1986.
Quinn, J. B., *The Strategy Process,* Englewood Cliffs, NJ: Prentice Hall, 1988.

2.3.1 Technical Entrepreneurship

A. D. Little, "New Technology-Based Firms in the United Kingdom and the Federal Republic of Germany," London: Wilton House, 1977.
Allison, G. T., *Essence of Decision: Explaining the Cuban Missile Crisis,* Boston, Little Brown & Co., 1971.
Bahrami, H. and Evans, H., "Stratocracy in High Technology Firms," *California Management Review,* Fall 1987, pp. 51–66.
Bahrami, H. and Evans, S., "*The Empiricist Mode of Strategy-Making in High Technology Firms,*" University of California Berkeley, Business and Public Policy Working Paper OBIR-28, Sept 1988.
Bollinger, L., Hope, K. and Utterback, J. M., "A Review of Literature and Hypotheses on New Technology-Based Firms," *Research Policy,* 12, 1, Feb. 1983.
Bourgeois, L. J. and Eisenhardt, K. M. "Strategic Decision Processes in Silicon Valley: The Anatomy of the Living Dead," *California Management Review,* Fall 1987, pp. 143–159.
Bourgeois, L. J., III and Eisenhardt, K. M., "Strategic Decision Processes in High Velocity Environments: Four Cases in the Microcomputer Industry," *Management Science,* 34, 7, July 1988, pp. 816–835.
Bower, J. L., *Managing the Resource Allocation Process,* Boston, Graduate School of Business Administration, Harvard University, 1970.
Bullock, M., *Academic Enterprise, Industrial Innovation and the Development of High Technology Financing in the United States,* London: Brand, 1983.
Burgelman, R. A., "On the Interplay of Process and Content in Internal Corporate Ventures: Action and Cognition in Strategy-Making," *Academy of Management Proceedings,* 1984, pp. 2–6.
Chandler, A. D., *Strategy and Structure,* Cambridge, MA: MIT Press, 1962.
Cooper, A. C., and Komives, J. L., eds, *Technical Entrepreneurship: A Symposium,* Milwaukee, Center for Venture Management, 1972.
Cooper, A. C., "Spin-Offs and Technical Entrepreneurship," *IEEE Transactions on Engineering Management,* EM-18, 1, February 1971, pp. 2–6.
Cooper, A. C., "Technical Entrepreneurship: What Do We Know?" *R&D Management,* 3, 2, February 1983.
Dorfman, N. S., "Massachusett's High Technology Boom in Perspective," *Center for Policy Alternatives,* M.I.T., 1982.
Eisenhardt, K. M. and Forbes, N., "Technical Entrepreneurship: An International Perspective," *Columbia Journal of World Business,* 19, 4, Winter 1984.
Gobeli, D. H. and Rudelis, W., "Managing Innovation: Lessons From the Cardiac Pacing Industry," *Sloan Management Review,* Summer 1985, pp. 29–43.
Hanan, M., *Venture Management,* New York: McGraw-Hill, 1976.

Herbert, R. F. and Link, A. N., *The Entrepreneur: Mainstream Views and Radical Critiques,* New York; Praeger, 1982.

Lindblom, C. E., "The Science of Muddling Through," *Public Administration Review,* Spring 1959, pp. 79–88.

Maidique, M. A., "Entrepreneurs, Champions, and Technological Innovation," *Sloan Management Review,* 21, 2, Winter 1980.

March, J. G., and Olsen, J. P., *Ambiguity and Choice in Organizations,* Norway, Universitetsforlaget, 1976.

Mintzberg, H. and McHugh, A., "Strategy Formation in an Adhocracy," *Administrative Science Quarterly,* 30, 1985, pp. 160–197.

Mintzberg, H. and Waters, J. A., "Of Strategies, Deliberate and Emergent," *Strategic Management Journal,* 6, 1985, pp. 257–272.

Mintzberg, H. and Waters, J. A., "Tracking Strategy in an Entrepreneurial Firm," *Academy of Management Journal,* 25, 3, 1982, pp. 465–499.

Mintzberg, H., "Crafting Strategy," *Harvard Business Review,* July–Aug. 1987.

Moore, J. R., "Unique Aspects of High Technology Enterprise Management," *IEEE Transactions on Engineering Management,* EM-23, 1, Feb 1975, pp. 10–20.

Pennings, J. M., "An Ecological Perspective on the Creation of Organizations," in J. R. Kimberly and R. H. Miles, eds., *The Organizational Life Cycle,* San Francisco: Jossey-Bass Publishers, 1980.

Penrose, E. T., *The Theory of the Growth of the Firm,* White Plains, NY: M. E. Sharpe, 1980.

Roberts, E. B. and Hauptman, O., "The Financing Threshold Effect on Success an Failure of Biomedical and Pharmaceutical Startups," *Management Science,* 33, 3, March 1987, pp. 381–394.

Roberts, E. B. and Wainer, H. A., "New Enterprises on Route 128," *Science Journal,* December 1968.

Roberts, E. B. and Wainer, H. A., "Some Characteristics of Technical Entrepreneurs," *IEEE Transactions on Engineering Management,* EM-18, 3, Aug. 1971, pp. 100–109.

Roberts, E. B., "A Basic Study of Innovations," *Research Management,* July 1968.

Roberts, E. B., "Entrepreneurship and Technology: A Basic Study of Innovators," *Research Management,* 11, 4, July 1968.

Romanelli, E., "New Venture Strategies in the Minicomputer Industry," *California Management Review,* Fall 1987, pp. 160–175.

Rothwell, R., "The Characteristics of Successful Innovators and Technically Progressive Firms," *R&D Management,* 7, 3, 1977, pp. 191–206.

Rothwell, R., "The Role of Small Firms in the Emergence of New Technologies," *Omega,* 12, 1, 1984, pp. 19–29.

Roure, J., *Success and Failure of High-Growth Technological Ventures: The Influence of Prefunding Factors,* Ph.D. dissertation, Dept. of Industrial Engineering and Engineering Management, Stanford University, 1987.

Rumelt, R. P., "Theory, Strategy and Entrepreneurship," in J. Teece, ed., *The Competitive Challenge,* Cambridge, MA: Ballinger, 1987.

Schumpeter, J. A., *The Theory of Economic Development,* Cambridge, MA: Harvard University Press, 1934.

Smilor, R. W., "Managing the Incubator System: Critical Success Factors to Accelerate

New Company Development,'' *IEEE Transactions on Engineering Management,* EM-34, 3, Aug. 1987, pp. 146–155.

Timmons, J. A., ''New Venture Creation: Models and Methodologies,'' in C. A. Kent, D. L. Sexton and K. H. Vesper, eds., *Encyclopedia of Entrepreneurship,* Englewood Cliffs, NJ: Prentice-Hall, 1982.

Vesper, K. H., *Frontiers of Entrepreneurship Research,* Wellesley, MA: Babson College, 1983.

2.3.2 The Technology Strategy Process in Larger Firms [see also 2.1.4.10]

Abell, P., *Organizations as Bargaining and Influence Systems,* London: Heinemann; New York, Halsted Press, 1975.

Adler, P. S. and Borys, B., ''A Guide for the Perplexed: A Meta-Theoretical Framework for the Study of Organizational Performance,'' Stanford University, April 1988.

Arrow, K. and Hurwicz, L., eds., *Studies in Resource Allocation Processes,* Cambridge: Cambridge U. P., 1977.

Bower, J. L., *Managing The Resource Allocation Process,* Boston, MA: Harvard Business School Press, 1970.

Business Week, ''TRW Leads a Revolution in Managing Technology,'' Nov. 15, 1983.

Doz, Y., Angelmar, R. and Prahalad, C. K., ''Technological Innovation and Interdependence: A Challenge for the Large Complex Firm,'' in M. Horwitch, ed., *Technology in Society,* 7, 2/3, 1986, pp. 14–34.

Eccles, R. G. *The Transfer Pricing Problem: A Theory for Practice,* Lexington, MA: Lexington Books, 1985.

Gupta, A. K. and Govandaranjan, V., ''Resource Sharing Among SBUs,'' *Academy of Management Journal,* 29, 4, 1986, pp. 695–714.

Horwitch, M. and Prahalad, C. K., ''Managing Technological Innovation—Three Ideal Modes,'' *Sloan Management Review,* Winter 1976.

Horwitch, M. and Thietart, R. A., ''The Effect of Business Interdependencies on Product R&D-Intensive Business Performance,'' *Management Science,* 33, 21, Feb. 1987, pp. 178–197, 2.1.4.9a.

Maidique, M. and Hayes, R. H., ''The Art of High Technology Management,'' *Sloan Management Review,* 25, Winter 1984, pp. 18–31.

''Managing Technology at Dow Chemical: Interview with Paul F. Oreffice,'' *Booz, Allen & Hamilton: Outlook,* Fall/Winter 1981.

March, J. G., ''The Business Firm as a Political Coalition,'' *Journal of Politics,* 24, 1962, pp. 662–678.

Mintzberg, H., *Power in and Around Organizations,* Englewood Cliffs, NJ: Prentice-Hall, 1983.

Mitchell, G. R., ''New Approaches for the Strategic Management of Technology,'' in M. Horwitch, ed., *Technology in Society,* 7, 2/3, 1986.

Murray, E. A. Jr., ''Strategic Choice as a Negotiated Outcome,'' *Management Science,* 24, 9, May 1978.

Ouchi, W. G., ''Markets, Bureaucracies and Clans,'' *Administrative Science Quarterly,* 25, 1, 1980.

Pettigrew, A., *The Politics of Organizational Decision Making,* London: Tavistock 1973.
Pfeffer, J. and G. R. Salancik, "Organizational Decision Making as a Political Process," *Administrative Science Quarterly,* 19, 1974, pp. 135–151.
Pondy, L. R., "Toward A Theory of Internal Resource Allocation," in M. N. Zald, ed., *Power in Organizations,* Nashville, Tenn.: Vanderbilt University Press, 1970, pp. 270–311.
Porter, M. E., *Competitive Advantage,* New York: Free Press, 1985, Chapter 5.
Quinn, J. B., "Managing Innovation: Controlled Chaos," *Harvard Business Review,* May–June 1985.
Quinn, J. B., "Technological Innovation, Entrepreneurship and Strategy," *Sloan Management Review,* Spring 1979.
Schoen, D. R., "Managing Technological Innovation," *Harvard Business Review,* May–June 1969.
Schon, D. A., "Champions for Radical New Inventions," *Harvard Business Review,* March–April 1963.
Yavitz, B. and Newman, W. H., *Strategy in Action: The Execution, Politics and Payoff of Business Planning,* New York: Free Press, 1982.

3. PROJECTS

3.1.1 The Innovation Process [see also 1.1]

Abend, C. J., "Innovation Management: The Missing Link in Productivity," *Management Review,* June 1979.
Albrecht, E. and Kant, H., "A Model of the Cycle Science-Technology-Production and its Application to the Development of Semiconductor Physics and Industry," *R&D Management,* 8, 1978, pp. 119–125.
Allen, T. J., *Managing the Flow of Technology,* Cambridge, MA: MIT Press, 1977.
Archibald, R. D., *Managing High-Technology Programs and Projects,* New York: John Wiley, 1976.
Baker, N. R. and Sweeney, D. J., "Toward a Conceptual Framework of the Process of Organized Innovation Within the Firm," *Research Policy,* 7, 1978, pp. 150–174.
Balderston, J., Birnbaum, P., Goodman, R. and Stahl, M., *Modern Management Technologies in Engineering and R&D,* New York: Van Nostrand, 1985.
Bessaret, J. R., "Influential Factors in Manufacturing Innovation," *Research Policy,* 11, 1982, pp. 117–132.
Bright, J. R., *Research, Development and Technological Innovation: An Introduction,* Homewood, ILL: Richard D. Irwin, 1964.
Burns, T. and Stalker, G. M., *The Management of Innovation,* London: Tavistock, 1961.
Carter, C. F. and Williams, B. R., *Industrial and Technological Progress: Factors Governing the Speed of Application of Science,* London: Oxford UP 1957.
Cetron, M. J., Goldhar, J. D. eds., *The Science of Managing Organized Technology,* New York: Gordon and Breach, 1971.
Chakrabarti, A. K., "The Role of Champions in Product Innovation," *California Management Review,* Winter 1974, pp. 58–62.
Clark, T. E., "Decision-Making in Technologically Based Organizations: A Literature

Survey of Present Practice,'' *IEEE Transactions on Engineering Management,* EM-21, 1, February 1974, pp. 9–23.

Conference Board, *Organization for New Product Development,* New York: The Conference Board, 1966.

Cooper, R. G., ''A Process Model for Industrial New Product Development,'' *IEEE Transactions on Engineering Management,* EM-30, 1, Feb. 1983, pp. 2–11.

Dean, B. V. ed., *Project Management,* Amsterdam; North-Holland 1985.

Ettlie, J. E., ''Performance Gap Theories of Innovation,'' *IEEE Transactions on Engineering Management,* EM-30, 2, May 1983, pp. 39–52.

Fischer, W. A., Hamilton, W., McLaughlin, C. P. and Zmud, R. W., ''The Elusive Product Champion,'' *Research Management,* May–June 1986, pp. 13–16.

Frohman, A. L., ''Critical Mid-Management Functions for Innovative R&D,'' *Research Management,* July 1976, 19, 4, pp. 7–13.

Gerstenfeld, A., *Effective Management of Research and Development,* Reading, MA: Addison-Wesley, 1970.

Ghee, S., ''Factors Affecting the Innovation Time-Period,'' *Research Management,* Jan. 1978, pp. 37–42.

Gilfillan, S. C., *The Sociology of Invention,* Cambridge, MA: MIT Press, 1970 (originally 1935).

Henwood, F. and Thomas, G., *Science, Technology and Innovation: A Research Bibliography,* New York: St. Martins Press, 1984.

Hollander, S., *The Sources of Increased Efficiency: A Case Study of Du Pont Rayon Manufacturing Plants,* Cambridge MA: MIT Press, 1965.

Horsmans, J. W., ''Innovation Management for an Industrial Product,'' *Research Policy,* 8, 3, 1979, pp. 274–283.

Jewkes, J. and others, *The Sources of Invention,* 2nd ed., London, MacMillan, 1969.

Katz, R. and Tushman, M., ''An Investigation into the Managerial Roles and Career Paths of Gatekeepers and Project Supervisors in a Major R&D Facility,'' *R&D Management,* 11, 3, 1981, pp. 103–110.

Kelly, P. and Kranzberg, M., eds., *Technological Innovation: A Critical Review of Current Knowledge,* San Francisco: San Francisco Press 1978.

Kline, S. J. and Rosenberg, N., ''An Overview of Innovation'' in R. Landau and N. Rosenberg, eds., *The Positive Sum Strategy,* Washington, DC: National Academy Press, 1986.

Lane, H. W., Beddows, R. G. and Lawrence, P. R., *Managing Large Research and Development Programs,* Albany: State University of New York Press, 1981.

Love, S. F., *Planning and Creating Successful Engineered Designs: Managing the Design Process,* North Hollywood, CA: Advanced Professional Development, 1986.

Mansfield, E., Rapoport, J., Schnee, J., Wagner, S. and Hamburger, M., *Research and Innovation in the Modern Corporation,* New York: Norton, 1971.

Mensch, G. O., ''Innovation Management in Diversified Corporations: Problems of Organization,'' *Human Systems Management,* 3, 1982, pp. 10–20.

Metcalfe, J. S., ''Impulse and Diffusion in the Study of Technological Change,'' *Futures,* 13, 5, 1981, pp. 347–359.

Meyer, A. C., ''The Flow of Technological Innovation in an R&D Department,'' *Research Policy,* 14, 6, Dec. 1985, pp. 315–328.

Mohr, L. B., "Determinants of Innovation in Organizations," *American Political Science Review,* 63, 1969, pp. 111–126.

Normann, R., "Organizational Innovativeness: Product Variation and Reorientation," *Administrative Science Quarterly,* 16, 1971.

Quinn, J. B., *Yardsticks of Industrial Research,* NY: Ronald Press, 1959.

Roberts, E. B. and Fusfield, A. R., "Staffing the Innovative Technology-Based Organization," *Sloan Management Review,* 22, 3, Spring 1981.

Roberts, E. B., "Generating Effective Corporate Innovation," *Technology Review,* Oct–Nov. 1977.

Roberts, E. B., "Managing Invention and Innovation," *Research Technology Management,* Jan–Feb. 1988, pp. 11–29.

Roberts, E. B., "Managing Invention and Innovation: What We've Learned," *Research Technology Management,* Jan–Feb. 1988, pp. 11–29.

Roberts, E. B., "Stimulating Technological Innovation—Organizational Approaches," *Research Management,* November 1979, pp. 26–30.

Rosenberg, N., "The Direction of Technological Change: Inducement Mechanisms and Focusing Devices," in *Perspectives on Technology,* New York: Cambridge UP, 1976.

Rosenbloom, R. S. and Cusumano, M. A., "Technological Pioneering: The Birth of the VCR Industry," *California Manugement Review,* 29, 4, Summer 1987, pp. 51–76.

Schon, D. A., "Champions for Radical New Inventions," *Harvard Business Review,* 41, 2, March–April 1963, pp. 77–86.

Schon, D. A., *Technology and Change,* Oxford: Pergamon, 1967.

Shrivastava, P. and Souder, W. E., "The Strategic Management of Technological Innovations: A Review and A Model," *Journal of Management Studies,* 24, 1, January 1987, pp. 25–41.

Sommers, W. P., "Improving Corporate Performance Through Better Management of Innovation," *Booz-Allen & Hamilton: Outlook,* Fall/Winter, 1981.

Souder, W. E., *Managing New Product Innovations,* Lexington, MA: Lexington, 1987.

Stobaugh, R., "Creating a Monopoly: Product Innovation in Petrochemicals," in R. S. Rosenbloom, ed., *Research on Technological Innovation, Management and Policy,* 2, 1985, pp. 81–112.

"The Role of the Product Champion: Conference Report," *R&D Management,* 15, 2, 1985, pp. 71–72.

Thompson, V. A., "Bureaucracy and Innovation," *Administrative Science Quarterly,* 10, June 1965.

Tornatzky, L. G. et al., *The Process of Technological Innovation: Reviewing the Literature,* Washington, DC: National Science Foundation, 1983.

Tushman, M. L. and Scantan, T. J., "Boundary Spanning Individuals: Their Roles in Information Transfer and their Antecedants," *Academy of Management Journal,* 24, 2, 1981, pp. 289–305.

Utterback, J. M., Hollomon, J. H., Sirbu, M. A., Jr. and Allen, T. J., "The Process of Innovation in Five Industries in Europe and Japan," *IEEE Transactions on Engineering Management,* EM-23, 1, Feb. 1975, pp. 3–9.

Utterback, J. M., "Innovation in Industry and Diffusion of Technology," *Science,* 83, Feb. 1984, pp. 620–626.

Utterback, J. M., "The Process of Technological Innovation within the Firm," *Academy of Management Journal,* 14, March 1971, pp. 75–88.

Van de Ven, A. H., "Central Problems in the Management of Innovation," *Management Science,* May 1986.

Von Hippel, E., "Appropriability of Innovation Benefit as a Predictor of the Source of Innovation," *Research Policy,* 11, 2, April 1982.

Wilson, J. Q., "Innovation in Organization: Notes Towards a Theory," in J. D. Thompson, ed. *Approaches to Organizational Design,* Pittsburgh: University of Pittsburgh Press, 1966.

3.1.2 Determinants of Innovation Success

Baker, B. N., Murphy, D. C. and Fisher, D., "Factors Affecting Project Success," in D. I. Cleland and W. R. King, eds., *Project Management Handbook,* New York: Van Nostrand, 1983.

Baker, N. R., Siegman, J. and Rubenstein, A. H., "The Effects of Perceived Needs for Industrial Research and Development Projects," *IEEE Transactions on Engineering Management,* EM-14, 4, Dec. 1967, pp. 156–162.

Cooper, R. G. and Kleinschmidt, E. J., "New Products: What Separates Winners and Losers?" *Journal of New Product Management,* 4, 3, Sept. 1987, pp. 169–184.

Cooper, R. G., "A Process Model for Industrial New Product Development," *IEEE Transactions on Engineering Management,* EM-30, 1, Feb. 1983, pp. 2–11.

Cooper, R. G., "Identifying Industrial New Product Success: Project New Prod," *Industrial Marketing Management,* 8, 1979, pp. 124–135.

Cooper, R. G., "Project New Prod: Factors in New Product Success," *European Journal of Marketing,* 14, 5/6, 1980, pp. 277–292.

Cooper, R. G., "The Dimensions of Industrial New Product Success and Failure," *Journal of Marketing,* 43, Summer 1979, pp. 93–107.

Crawford, C. M., "New Product Failure Rates: A Reprise," *Research Management,* July–Aug. 1987, pp. 20–24.

Gerstenfeld, A., "A Study of Successful Projects, Unsuccessful Projects and Projects in Process in West Germany," *IEEE Transactions on Engineering Management,* EM-23, 3, 1976.

Globe, S., Levy, G. W. and Schwartz, C. M., "Key Factors and Events in the Innovation Process," *Research Management,* July 1973.

Lilien, G. and Yoon, E., "*Determinants of New Industrial Product Performance: A Strategic Re-examination of the Empirical Literature,*" Institute for the Study of Business Markets Report 4-1986, Pennsylvania State University, 1986.

Maidique, M. A. and Zirger, B. J., "A Study of Success and Failure in Product Innovation: The Case of the U.S.," *IEEE Transactions on Engineering Management,* EM-31, 4, Nov. 1984, pp. 192–203.

McIntyre, S. H., "Obstacles to Corporate Innovation," *Business Horizons,* Jan–Feb. 1982.

Might, R. J., "The Role of Structural Factors in Determining Project Management Success," *IEEE Transactions on Engineering Management,* EM-32, 2, May 1985, pp. 71–77.

Myers, S. and Marquis, D., *Successful Industrial Innovations*, Washington, DC: National Science Foundation, 1969.

Pavitt, K., *The Conditions of Success in Technological Innovation*, Paris: OECD, 1971.

Pinto, J. K. and Slevin, D. P., "Critical Factors in Successful Project Implementation," *IEEE Transactions on Engineering Management*, EM-34, 1, Feb. 1987, pp. 22–27.

Rothwell, R., Freeman, C., Horsley, A., Jervis, V. T. P., Robertson, A. B. and Townsend, J., "SAPPHO Updated—Project SAPPHO, Phase II," *Research Policy*, 3, 1974, pp. 258–291.

Rothwell, R., "The Characteristics of Successful Innovators and Technically Progressive Firms," *R&D Management*, 7, 3, 1977, pp. 191–206.

Rubenstein, A. H., Chakrabarti, A. K., O'Keefe, R. D., Souder, W. E. and Young, H. C., "Factors Influencing Innovation Success at the Project Level," *Research Management*, 19, 13, May 1976, pp. 15–20.

Science Policy Research Unit, "Success and Failure in Industrial Innovation—Report on Project Sappho," London: CSII.

Souder, W. E. and Chakrabarti, A., "Industrial Innovations: A Demographical Analysis," *IEEE Transactions on Engineering Management*, EM-26, 4, November 1979, pp. 101–109.

3.1.3 Information Flows in Innovation

Allen, T. J. and Fusfield, A. R., "Design for Communication in the Research and Development Lab," *Technology Review*, May 1976.

Allen, T. J., and Cohen, S., "Information Flow in R&D Labs," *Administrative Science Quarterly*, 14, 1969, pp. 12–19.

Allen, T. J., *Managing the Flow of Technology*, Cambridge, MA: MIT Press 1977.

Allen, T. J., "Performance of Information Channels in the Transfer of Technology," *Industrial Management Review*, 8, 1966.

Allen, T. J., "Studies of the Problem-Solving Process in Engineering Design," *IEEE Transactions on Engineering Management*, EM-13, 2, June 1960, pp. 72–82.

Aloni, M., "Patterns of Information Transfer Among Engineers and Applied Scientists in Complex Organizations," *Scientometrics*, 8, 5–6, 1985, pp. 279–300.

Baker, N. and Freeland, J. R., "Structuring Information Flow to Enhance Innovation," *Management Science*, 19, 1, Sept. 1972, pp. 105–116.

Crane, D., *Invisible Colleges*, Chicago, IL: University of Chicago Press 1972.

Czepiel, J. A., "Patterns of Interorganizational Communications and the Diffusion of a Major Technological Innovation in a Competitive Industrial Community," *Academy of Management Journal*, 18, 1, March 1978.

Frost, R. and Whitley, R., "Communication Patterns in a Research Lab," *R&D Management*, 1, 1971, pp. 71–79.

Golhar, J. D., Bragaw, L. K. and Schwartz, J. J., "Information Flows, Management Styles, and Technological Innovation," *IEEE Transactions on Engineering Management*, EM-23, 1, Feb. 1975, pp. 51–61.

Katz, R. and Tushman, M. L., "Communication Patterns, Project Performance, and Task Characteristics: An Empirical Evaluation and Integration in an R&D Setting," *Organizational Behavior and Human Performance*, 1979, 23, pp. 139–162.

Keller, R. and Holland, W., "Boundary Spanning Roles in R&D Organization," *Academy of Management Journal,* 1975, 18, pp. 388–393.

Keller, R. and Holland, W., "Individual Characteristics of Innovativeness and Communication in Research and Development Organizations," *Journal of Applied Psychology,* 1978, 63, pp. 759–762.

Myers, S. and Marquis, D. G. *Successful Industrial Innovations,* Washington: National Science Foundation, 1969.

Rosenbloom, R. S. and Wolek, F. W., *Technology and Information Transfer,* Boston, MA: Harvard University Graduate School of Business Administration, 1970.

Tushman, M. L. and Nadler, D. A., "Communication and Technical Roles in R&D Laboratories: An Information Processing Approach," *TIMS Studies in the Management Science,* 15, 1980, pp. 91–112.

Tushman, M. L. and Scantan, T. J., "Boundary Spanning Individuals: Their Roles in Information Transfer and their Antecedants," *Academy of Management Journal,* 24, 2, 1981, pp. 289–305.

Tushman, M. L., "Communication Across Organizational Boundaries: Special Boundary Roles in the Innovation Process," *Administrative Science Quarterly,* 22, 1977, pp. 587–605.

Tushman, M. L., "Technical Communication in Research and Development Laboratories: Impact of Project Work Characteristics," *Academy of Management Journal,* 22, 1979, pp. 624–645.

Utterback, J. M., "The Process of Innovation: A Study of the Origination and Development of Ideas for New Scientific Instruments," *IEEE Transactions on Engineering Management,* EM-18, 4, Nov. 1971, pp. 124–131.

Von Hippel, E., "The Dominant Role of Users in the Scientific Instrument Innovation Process," *Research Policy,* 5, 3, 1976, pp. 212–239.

Westood, A. R., "R&D Linkages in a Multi-Industry Corporation," *Research Management,* May–June 1984, pp. 23–26.

3.1.4 Diffusion of Innovations

Abraham, C. and Hayward, G., "Towards a Microscopic Analysis of Industrial Innovations: From Diffusion Curves to Technological Integration Through Participative Management," *Technovation,* 3, 1985, pp. 3–17.

Antonelli, C., "The International Diffusion of New Information Technologies," *Research Policy,* 15, 3, June 1986.

Bessant, J. R., "Influential Factors in Manufacturing Innovation," *Research Policy,* 11, 2, 1982, pp. 117–132.

Brown, L. A., *Innovation Diffusion: A New Perspective,* London; Methuen, 1981.

Cohen, W. M. and Levin, R. C., "Empirical Studies of Innovation and Market Structure," in R. Schmalensee and R. Willig, eds., *The North-Holland Handbook of Industrial Organization,* New York: North-Holland, 1988, forthcoming.

David, P. A., " A Contribution to the Theory of Diffusion," Center for Research in Economic Growth, Memo No. 71, Stanford University, 1969.

David, P. A., "Technological Diffusion, Public Policy and Industrial Competitiveness," in R. Landau and N. Rosenberg eds., *The Positive Sum Strategy,* Washington, DC: National Academy, 1986.

Davies, S., *The Diffusion of Process Innovations,* Cambridge: Cambridge University Press, 1979.

Gold, B., "On the Adoption of Technological in Industry—Superficial Models and Complex Decision Processes," *Omega,* 8, 5, 1980, pp. 505–516.

Gold, B., "Technological Diffusion in Industry—Research Needs and Shortcomings," *Journal of Industrial Economics,* March 1981, pp. 247–269.

Hayward, G., Allen, D. and Masterson, J., "Characteristics and Diffusion of Technological Innovations," *R&D Management,* 7, 1, 1976.

Katz, E., Levin, M. L. and Hamilton, H., "Traditions of Research on the Diffusion of Innovations," *American Sociological Review,* April 1963.

Mansfield, E., *Industrial Research and Technological Innovation,* New York: W. W. Norton, 1968.

Mansfield, E., "Technical Change and the Rate of Imitation," *Econometrica,* 29, 1961, pp. 741–766.

Mansfield, E., *The Economics of Technological Change,* New York: W. W. Norton, 1968.

Mansfield, E., "The Speed of Response of Firms to New Techniques," *Quarterly Journal of Economics,* 77, 1963, pp. 290–311.

Nabseth, L. and Ray, G., *The Diffusion of New Industrial Processes,* Cambridge: UP, 1974.

Rogers, E. M. and Eveland, J. D., "Diffusion of Innovations Perspectives on National R&D Assessment: Communication and Innovation in Organizations," in P. Kelly and M. Kranzberg, eds., *Technological Innovation: A Critical Review of Current Knowledge,* San Francisco: San Francisco Press, 1978.

Rogers, E. M. and Floyd, F., *Communication of Innovations: A Cross-Culture Approach,* Shoemaker, 2nd ed., London: Collier-MacMillan, 1971.

Rogers, E. M. and Schoemaker, F., *Communication of Innovations,* New York: Free Press 1971.

Rogers, E. M., *Diffusion of Innovations,* New York Free Press, 1983.

Rosenberg, N., "Factors Affecting the Diffusion of Technology," in *Perspectives on Technology,* New York: Cambridge University Press, 1976.

Tilton, J. E., *International Diffusion of Technology: The Case of Semiconductors,* Washington, DC: Brookings, 1971.

Utterback, J. M., "Innovation in Industry and the Diffusion of Technology," *Science,* 183, Feb 1974, pp. 620–628.

3.1.5 Technology Transfer

Beaumont, C., Dingle, J. and Reithinger, A., "Technology Transfer and Applications," *R&D Management,* 11, 4, 1981.

Cetron, H. D. and Goldhar, J., *Technology Transfer,* Leiden: Noordhoff International, 1974.

Chakrabarti, A. K. and Rubenstein, A. H., "Interorganizational Transfer of Technology: Adoption of NASA Innovations," *IEEE on Transactions on Engineering Management,* EM-23, 1, Feb. 1975, pp. 20–34.

Chakrabarti, A. K., "Some Concepts of Technology Transfer: Adoption of Innovations in Organizational Context," *R&D Management,* 3, 3, 1973.

Dalman, C. J. and Westphal, L. E., "The Meaning of Technological Mastery in Relation to Transfer of Technology," *AAPS Annals,* Nov. 1981, pp. 12–26.

Dembo, V., "Technology Transfer Planning," *R&D Management,* 9, 3, 1979.

Douds, C. F., "The State of the Art in the Study of Technology Transfer: A Brief Survey," *R&D Management,* 1, June 1971.

Fischer, W., "Empirical Approaches to Understanding Technology Transfer," *R&D Management,* 6, 1976.

Foster, R. N., "Organize for Technology Transfer," *Harvard Business Review,* Nov–Dec. 1971.

Frosch, R., "R&D Choices and Technology Transfer," *Research Management,* May–June 1984.

Gartner, J. and Naiman, C., "Making Technology Transfer Happen," *Research Management,* May 1978.

Gartner, J. and Naiman, C., "Overcoming the Barriers to Technology Transfer," *Research Management,* March 1976.

Giberson, W. E., "Management of Technology Transfer in an Advanced Project: The Case of Surveyor," *IEEE Transactions on Engineering Management,* EM-16, 3, Aug. 1969, pp. 125–129.

Gruber, W. A. and Marquis, D. G., *Factors in the Transfer of Technology,* Cambridge, MA: MIT Press, 1969.

Hawkins, W. M., "Technology Transfer Programs at Lockheed," *IEEE Transactions on Engineering Management,* EM-16, 3, Aug. 1969, pp. 121–124.

Jervis, P., "Innovation and Technology Transfer—The Roles and Characteristics of Individuals," *IEEE Transactions on Engineering Management,* EM-22, 1, Feb. 1975, pp. 19–26.

Joshi, B., "International Transfer of Technology System" *IEEE Transactions on Engineering Management,* EM-24, 3, Aug. 1977, pp. 86–93.

Lambright, W. H. and Teich, A. H., "Technology Transfer as a Problem in Interorganizational Relationships," *Administration and Society,* 8, 1, May 1976.

Lester, M., "The Transfer of Managerial and Technological Skills by Electronic-Assembly Companies in Export-Processing Zones in Malaysia," in D. Sahal ed., *The Transfer and Utilization of Technical Knowledge,* Lexington, MA: Lexington, 1982.

Moore, J. R., "The Technology Transfer Process Between a Large Science-Oriented and a Large Market-Oriented Company—The North American Rockwell Challenge," *IEEE Transactions on Engineering Management,* EM-16, 3, Aug. 1969, pp. 111–115.

Quinn, J. B., "Technology Transfer by Multinational Companies," *Harvard Business Review,* Nov–Dec. 1969.

Richardson, J., ed., *Integrated Technology Transfer,* Mount Airey, Maryland: Lomond Books, 1979, p. 162.

Robbins, M. O. and Milliken, J. G., "Technology Transfer and the Process of Technological Innovation: New Concepts, New Models," *R&D Management,* 6, 1976.

Roberts, E. B. and Hauptman, O., "The Process of Technology Transfer to the New Biomedical and Pharmaceutical Firm," *Research Policy,* 15, 3, June 1986, pp. 107–120.

Rosenbloom, R. S., *Technology Transfer—Process and Policy: An Analysis of the Utilization of Technological By-products of Military and Space R&D,* Washington, National Planning Association, 1965, Special Report No. 62.

Rothwell, R., "Some Problems of Technology Transfer into Industry: Example from the

Textile Machinery Sector," *IEEE Transactions on Engineering Management,* EM-25, 1, Feb. 1978, pp. 15–19.
Sahal, D., "The Form of Technology Governs the Scope of its Transfer," in D. Sahal ed., *The Transfer and Utilization of Technical Knowledge,* Lexington, MA: Lexington, 1982.
Scheirer, M., "Approaches to the Study of Implementation," *IEEE Transactions on Engineering Management,* EM-30, 2, May 1983, pp. 76–82.
Stobaugh, R. and Wells Jr., L. T. eds., *Technology Crossing Borders,* Boston: Harvard Business School Press, 1984.
Teece, D. J., "Technology Transfer by Multinational Firms: The Resource Cost of Transferring Technological Know-How," *The Economic Journal,* June 1977, pp. 242–261.

3.2. Strategic and Operating Issues in Managing Technical Projects

Fujimura, J. H., "Constructing 'Do-able' Problems in Cancer Research: Articulating Alignment," *Social Studies of Science,* 17, 1987, pp. 257–293.
Giddens, A., *Central Problems in Social Theory,* London: MacMillan, 1979.
Imai, K., Nonaka, I. and Takeuchi, H., "Managing the New Product Development Process: How Japanese Companies Learn and Unlearn," in K. B. Clark, R. H. Hayes and C. Lorenz, eds., *The Uneasy Alliance: Managing the Productivity-Technology Dilema,* Boston, MA: Harvard Business School Press, 1985.
Isenberg, D. J., "The Tactics of Strategic Opportunism," *Harvard Business Review,* March–April 1987, pp. 92–97.
Jaikumar, R., "Postindustrial Manufacturing," *Harvard Business Review,* Nov–Dec. 1986, pp. 69–76.
Kelly, P. and Kranzberg, M. eds, *Technological Innovation: A Critical Review of Current Knowledge,* San Francisco: San Francisco Press 1978.
Kotter, J. P., *The Leadership Factor,* New York: Free Press, 1988.
Mintzberg, H., "the Manager's Job: Folklore and Fact," *Harvard Business Review,* July–Aug. 1975, pp. 49–61.
Mintzberg, H., *The Nature of Managerial Work,* New York: Harper & Row, 1973.
Pelz, D. C., "Quantitative Case Histories of Urban Innovations: Are There Innovating Stages?," *IEEE Transactions on Engineering Management,* EM-30, 2, May 1983, pp. 60–67.
Tornatzky, L. G. et al., *The Process of Technological Innovation: Reviewing the Literature,* Washington, DC: National Science Foundation, 1983.

3.2.1 Pre-project Phase

Abernathy, W. J. and Baloff, N., "Interfunctional Planning for New Product Introduction," *Sloan Management Review,* Winter 1972–73, pp. 25–43.
Abernathy, W. J. and Utterback, J. M., "Patterns of Industrial Innovation," in M. L. Tushman and W. L. Moore, eds., *Readings in the Management of Innovation,* Marshfield, MA: Pitman, 1982.

Bower, J. L., *Managing The Resource Allocation Process,* Boston, MA: Harvard Business School Press, 1970.

Burgelman, R. A., "A Process Model of Internal Corporate Venturing in the Diversified Major Firms," *Administrative Science Quarterly,* 28, 1983, pp. 223–244.

Cooper, R. G., "How New Product Strategies Impact on Performance," *Journal Product Innovation Management,* 1, 1984a.

Cooper, R. G., "The Strategy-Performance Link in Product Innovation," *R&D Management,* 14, 4, 1984b.

Gold, B., *Explorations in Managerial Economics: Productivity, Costs, Technology and Growth,* London: MacMillan, 1971.

Gold, B., "Technological Diffusion in Industry—Research Needs and Shortcomings," *Journal of Industrial Economics,* March 1981, pp. 247–269.

Hayes, R. H., Wheelwright, S. C. and Clark, K. B., *Dynamic Manufacturing,* New York: Wiley, 1988.

Johnson, S. C. and Jones, C., "How to Organize for New Products," *Harvard Business Review,* May–June 1957, pp. 49–62.

Wheelwright, S. C., "Product Development and Manufacturing Start-Up," *Booz-Allen & Hamilton: Manufacturing Issues,* 1985.

3.2.2 Post-project Phase

Argyris, C. and Schon, D., *Organizational Learning,* Reading: MA: Addison-Wesley, 1978.

Burgelman, R. A. "Strategy-Making as a Social Learning Process: The Case of Internal Corporate Venturing," *Interfaces,* 18, 3, May–June 1988, pp. 74–85.

Collier, D. W. and Gee, R. E., "A Simple Approach to Post-Evaluation of Research," *Research Management,* May 1973, pp. 12–17.

Fiol, C. M. and Lyles, M., "Organizational Learning," *Academy of Management Review,* 10, 4, 1985, pp. 803–813.

Gulliver, F. R., "Post-Project Appraisals Pay," *Harvard Business Review,* March–April 1987, pp. 128–133.

Hayes, R. H., Wheelwright, S. C. and Clark, K. B., *Dynamic Manufacturing,* New York: Wiley, 1988.

Hedberg, B. L. T., "How Organizations Learn and Unlearn," in N. C. Nystom and W. H. Starbuck, eds., *Handbook of Organizational Design,* Oxford: Oxford U.P., 1981, 1, pp. 3–27.

Imai, M., *Kaizen: The Key to Japanese Competitive Success,* New York: Random House, 1986.

Levinthal, D. A. and March, J. G., "A Model of Adaptive Organizational Search," *Journal of Economic Behavior and Organization,* 2, 1981, pp. 307–333.

Levinthal, D. A. and Yao, D. A., "The Search for Excellence: Organizational Inertia and Adaptation," *Management Science,* forthcoming.

Levitt, B. and March, J. G., "Organizational Learning," *Annual Review of Sociology,* 14, 1988.

Maidique, M. A., Zirger, B. J., "The New Products Learning Cycle," *Research Policy,* 14, 6, 1985, pp. 299–313.

Porter, J. G., Jr., "Post-Audits—An Aid to Research Planning," *Research Management*, Jan. 1978, pp. 28–30.

Weick, K. E., *The Social Psychology of Organizing*, Reading, MA: Addison-Wesley, 1979.

Wheelwright, S. C., "Product Development and Manufacturing Start-Up," *Booz-Allen & Hamilton: Manufacturing Issues*, 1985.

3.3 Strategic Lessons

Abernathy, W. J. and Utterback, J. M., "Patterns of Industrial Innovation," in M. L. Tushman and W. L. Moore, eds., *Readings in the Management of Innovation*, Marshfield, MA: Pitman, 1982.

Abetti, P., "Milestones for Managing Technological Innovations," *Planning Review*, 13, 2, March 1985, pp. 18–45.

Allen, T. J., Lee, D. M. S. and Tushman, M. L., "R&D Performance as a Function of Internal Communication, Project Management, and the Nature of the Work," *IEEE Transactions on Engineering Management*, EM-27, 1, 1980, pp. 2–12.

Allen, T. J., Tushman, M. L., Lee, D. M. S., "Technology Transfer as a Function of Position on Research, Development and Technical Services Continuum," *Academy of Management Journal*, 22, 4, 1979, p. 694–708.

Blandin, J. S. and Brown, W. B., "Uncertainty and Management's Search for Information," *IEEE Transactions on Engineering Management*, EM-24, 4, Nov 1977, pp. 114–119.

Clark, K. B. and Fujimoto, T., *"Overlapping Problem-Solving in Product Development*," Boston, MA: Harvard Business School Working Paper 87-048, 1987.

Cleland, D. I. and King, W. R., *Systems Analysis and Project Management*, 2nd ed. New York: McGraw-Hill, 1975.

Cleland, D. I. and Kocaoglu, D. F., *Engineering Management*, New York: McGraw-Hill, 1981.

Cooper, R. G., "The New Product Process: An Empirically-Based Classification Scheme," *R&D Management*, 16, 1, 1983.

Daft, R. L. and Lengel, R. H., "Organizational Information Requirements: Media Richness and Structural Design," *Management Science*, 32, 5, May 1986, pp. 554–571.

Dean, J. W., Jr., *Deciding to Innovate: How Firms Justify Advanced Technology*, Cambridge, MA: Ballinger, 1987.

Dickson, J. R., "Project Management: A New Agenda" in M. C. Grool, C. Visser, W. J. Vriethoff and G. Wijnen eds., *Project Management in Progress: Tools and Strategies for the 90s*, Amsterdam: North-Holland, 1986.

Downs, G. W., Jr. and Mohr, L. B., "Conceptual Issues in the Study of Innovation," *Administrative Science Quarterly*, 21, 1976, pp. 700–714.

Ettlie, J. E., Bridges, W. P. and O'Keefe, R. D., "Organization Strategy and Structural Differences for Radical Versus Incremental Innovation," *Management Science*, 30, 6, June 1984, pp. 682–695.

Goodman, P. S., "Impact of Task and Technology on Group Performance," in P. S. Goodman and Associates, *Designing Effective Work Groups*, San Francisco: Jossey-Bass, 1986.

Hauptman, O., *"The Different Roles of Communication in Software Development and Hardware R&D: Phenomenological Paradox or A-Theoretical Empiricism?"* Harvard Business School Working Paper 89-028, Nov 1988.

Jaikumar, R. and Bohn, R. E., "The Development of Intelligent Systems for Industrial Use: A Conceptual Framework," in R. S. Rosenbloom, ed., *Research on Technological Innovation, Management and Policy,* 3, 1986, pp. 169–212.

Katz, R. and Tushman, M., "An Investigation into the Managerial Roles and Career Paths of Gatekeepers and Project Supervisors in a Major Research Facility," *R&D Management,* 11, 31, 1981, pp. 103–110.

Katz, R. and Tushman, M., "Communication Patterns, Project Performance, and Task Characteristics: An Empirical Evaluation and Integration in an R&D Setting," *Organizational Behavior and Human Performance,* 23, 1979, pp. 139–162.

Keller, R. T. and Holland, W. E., "Boundary-Spanning Activity and Research and Development Management: A Comparative Study," *IEEE Transactions on Engineering Management,* EM-22, 4, Nov 1975, pp. 130–133.

Kelly, P. and Kranzberg, M., eds., *Technological Innovation: A Critical Review of Current Knowledge,* San Francisco: San Francisco Press 1978.

Kerr, S. and Von Glinow, M. A., "Issues in the Study of Professionals in Organizations: the Case of Scientists and Engineers," *Organizational Behavior and Human Performance,* 18, 1977, pp. 329–345.

Kimberly, J. R. and Evanisko, M. J., "Organizational Innovation: The Influence of Individual, Organizational and Contextual Factors on Hospital Adoption of Technological and Administrative Innovations," *Academy of Management Journal,* 24, 4, 1981, pp. 687–713.

Leifer, R. and Triscari, T., Jr., "Research Versus Development: Differences and Similarities," *IEEE Transactions on Engineering Management,* EM-34, 2, May 1987, pp. 71–78.

Levitt, B. and March, J. G., "Organizational learning," *Annual Review of Sociology,* 14, 1988.

Maidique, M. A. and Zirger, B. J., "New Products Learning Cycle," *Research Policy,* December 1985, pp. 1–40.

Marschak, T. A., Glenn, T. K. Jr. and Summers, R., *Strategy for R&D: Studies in the Microeconomics of Development,* New York: Springer-Verlag, 1967.

Moore, W. L. and Tushman, M. L., "Managing Innovation over the Product Life Cycle," in M. L. Tushman and W. L. Moore, eds., *Readings in the Management of Innovation,* Marshfield, MA: Pitman, 1982.

Radnor, M. and Rich, R. F., "Organizational Aspects of R&D Management: A Goal-Directed Contextual Perspective," *TIMS Studies in the Management Sciences,* 15, 1980, pp. 113–133.

Roberts, E. B., "Toward A New Theory for Research and Development," *Industrial Management Review,* 4, 1, Fall 1962, pp. 29–40.

Rogers, E. M. and Eveland, J. D., "Diffusion of Innovations Perspectives on National R&D Assessment: Communication and Innovation in Organizations," in P. Kelly and M. Kranzberg, eds., *Technological Innovation: A Critical Review of Current Knowledge,* San Francisco: San Francisco Press, 1978.

Rogers, E. M., *Diffusion of Innovations,* New York: Free Press, 1983.

Rosenbloom, R. S. and Wolek, F. W., *Technology and Information Transfer,* Boston, MA: Harvard University Graduate School of Business Administration, 1970.

Rouse, W. B. and Boff, K. R., *System Design: Behavioral Perspectives on Designers, Tools and Organizations,* New York: North Holland, 1987.

Small, H. and Grittin, B. C., "The Structure of Scientific Literatures, Part 1: Identifying and Graphing Specialities," *Science Studies,* 4, 1974, pp. 17–40.

Subramanyam, K., "Collaborative Publication and Research in Computer Science," *IEEE Transactions on Engineering Management,* EM-30, 4, Nov 1983, pp. 228–30.

Tornatzky, L. G. et al., *The Process of Technological Innovation: Reviewing the Literature,* Washington, DC: National Science Foundation, 1983.

Tushman, M. L., "Special Boundary Roles in the Innovation Process," *Administrative Science Quarterly,* 22, 1977, pp. 587–605.

Tushman, M. L., "Technical Communication in R&D Laboratories: The Impact of Project Work Characteristics," *Academy of Management Journal,* 21, 4, 1978, pp. 624–645.

Tushman, M. L., "Work Characteristics and Subunit Communication Structure: A Contingency Analysis," *Administrative Science Quarterly,* 24, 1979b, pp. 82–98.

Tushman, M. L., "Managing Communication Networks in R&D Laboratories," *Sloan Management Review,* 20, 2, Winter 1979a, pp. 27–49.

Twiss, B., *Managing Technological Innovation,* 2nd ed., London: Longman Group, 1980.

White, J. R., "Strategic Factors in Project-Based Business," in A. J. Kelly, ed., *New Dimensions in Project Management,* Lexington, MA: Lexington, 1982.

White, W., "A Risk/Action Model for the Differentiation of R&D Profiles," *IEEE Transactions on Engineering Management,* EM-29, 3, Aug 1982, pp. 88–93.

4. CONCLUSION

Kantrow, A. M., "The Strategy-Technology Connection," *Harvard Business Review,* July–August 1980.

National Research Council, *Management of Technology: The Hidden Competitive Advantage,* Washington, D.C.: National Academy Press, 1987.

TECHNOLOGY, VARIETY AND ORGANIZATION:

A SYSTEMATIC PERSPECTIVE ON THE COMPETITIVE PROCESS

J. S. Metcalfe and M. Gibbons

I. PERSPECTIVES

Our objective in this chapter is to outline a framework for the analysis of the relationship between technology and long-run competitive performance. We shall suggest that one promising strand of development involves adapting concepts and insights from evolutionary theory to generate a framework for the analysis of technological change. Within this framework (McElvey, 1982; Nelson and Winter, 1983; Silverberg, 1985;

Research on Technological Innovation, Management and Policy
Volume 4, pages 153–193

ISBN: 0-89232-798-7

Van Parijs, 1981) one is seeking to understand how technological change acts as a motor of structural change and economic development. Here, two issues are predominant: how new technological forms are created; and how these technological forms come to acquire economic significance. Both issues are at the root of any account of historical, structural change. Moreover, as far as the impacts of new technology are concerned, on employment or regional development, it is the economic significance, or weight of the technology that is the crucial issue for any explanation of structural change. It is because these two issues are, in effect, ones of variety and selection that the evolutionary metaphor becomes appropriate. Central to this endeavor is the need to distinguish sharply between selection environments, organizations, and technologies. It is on this triad that this short essay is focused.

There can be no doubt from the historical record that technological advance is associated with rapid changes in economic structures, and in the relative significance of particular technologies and firms. The microelectronics and consumer electronics industry are just two sectors of many which have been implicated in such fundamental economic change on an international scale since 1960. (Braun and MacDonald, 1978; Malerba, 1985; Rosenbloom and Abernathy, 1985; Millstein and Borrus et al. 1983). The view we take in this chapter is that these changes are the outcome of a competitive process that is evolutionary in two senses: it entails selection across rival alternatives; and it entails persistent elements of incremental, cumulative, path dependent change in those technological alternatives. Changes which are punctuated at intervals by major kaleidoscopic jumps, or saltations, in the set of available technologies.

In our approach we find it helpful to analyze technology at two conceptual levels. In terms of artifacts, the products and processes of production that firms reveal in the market place; and in terms of the corresponding knowledge bases, the ideas, concepts and modes of enquiry that are necessary to generate a particular revealed performance (Layton, 1974). Bridging the two dimensions of technology is the firm, that organization which articulates a knowledge base to design and implement a particular level of revealed performance. One immediate implication of this is that both the knowledge base and revealed technological performance are concepts inseparable from questions of organizational structure and activity. But paradoxically perhaps, the competitive environment does not select directly with respect to organizations or knowledge bases, but rather with respect to the products they underpin and the corresponding methods of production. For the firm, its technological knowledge base constitutes a major invisible asset along with its knowledge of its markets and its organization (Itami, 1987). For present purposes, however, it is the intellectual, scientific and technological knowledge base that is the focus of our attention.

The underlying presumption in all that follows is that economic change is driven by variety in economic performance, which in turn is contingent upon variety in technology or in organizational form. We wish to explore why it is that firms exhibit differences in revealed technological performance, in creativity, and in the organization of knowledge resources; and why these differences change over time.

As such, the evolutionary framework to be developed is more concerned with frequencies of events and phenomena than with ideal, representative types and there is a considerable shift in intellectual orientation in this change of emphasis. More is at stake here than epistemology. The shift from analyzing ideal cases to examining frequencies and their distribution, is central to the elaboration of an evolutionary perspective of the sort we are proposing. The shift from classical to distributional modes of explanation has occurred in biology in terms of the shift from typological to population thinking about species (Mayr, 1982; Sober, 1985). In typological thinking species are regarded as fixed and identifiable in terms of a few distinct characteristics which represent the essence of the entity. In this view all variations around the ideal type are accidental and, in interpretational terms, they are aberrations.

By contrast, in population thinking, species are described in terms of a distribution of characteristics and whereas, in typological thinking, variation is a nuisance, in population thinking it is of all-consuming interest because it is the variety in the system which drives the evolutionary process. Moreover, the changes over time in statistical moments derived from the characteristics distribution are an index of the rate and direction of evolutionary change. Such modes of reasoning have been successfully employed by Hannan and Freeman (1977) and McElvey (1982) in their discussions of organizational change.

Before moving on to the substance of our argument, it will be as well to remind ourselves here of the essential mechanisms of evolutionary change. These are: the principle of variation, that members of the population vary with respect to at least one characteristic with selective significance; the principle of heredity, that there is continuity over time in the form of the species under investigation; and, the principle of selection, that some forms are better fitted to environmental pressure and thus increase in relative significance compared to inferior forms.

To transfer these concepts uncritically to a social science context is correctly recognized to be untenable although no less tenable than is the widespread appeal to mechanical concepts in economics in particular. Nonetheless, applied carefully to the context of technological competition they prove to be remarkably fruitful, primarily because they are ideally suited to cope with two enduring historical facts, namely variety and change. As authors from Marx and Schumpeter have recognized, innovation enhances variety. Marshall (1920) it will be remembered, argued that variation was the chief form of progress (*Principles,* p. 355) and suggested that the institutional structures of an industry have a major impact on the range and rate of variety enhancing experimentation which it undertakes (Loasby, 1982). By contrast, the related phenomena of imitation and technology transfer diminish variety. Economic environments then provide the basis for selection between competing technologies, by establishing price structures that provide a direct evaluation of the performance characteristics of rival products and processes. While products and processes are the direct units of selection this necessarily entails indirect selection across the firms that articulate those technologies. But the two levels of selection must be kept quite distinct, the selection and survival of firms involves considerations beyond those which determine the selection and survival of technologies. Paraphrasing Sober (1985, p. 100), there is selection 'for' performance characteristics and selection 'of' technologies and by implication firms.

A central element in this process of selection and evaluation of technologies is that it determines the relative profitability of different lines of production. The profits which firms thereby acquire provide the basis for investment and marketing activities that are, in turn, the basis for further change of the relative economic significance of these firms. More profitable technologies have the potential to displace less profitable technologies, but whether this potential is realized depends crucially on the behavior of the articulating organizations. The distribution of profitability across technologies and firms has another major evolutionary impact. Insofar as technological change is resource dependent, profits provide the basis for advancing technology so that the terms of technological competition are continually redefined. Whatever one's views on the role of the price mechanism in equilibrating markets, the fundamental fact remains that the structure of prices in a selection environment is a major determinant of the distribution of profits between the competing firms and thus of the distribution of resources to engage in rivalrous behaviour. Thus major elements of change become endogenous to the competitive process. In changing the relative significance of competing technologies, selection also results in changes in the price structures that evaluate performance characteristics, so reshaping the selection environment. Indeed one of the central themes of the evolutionary approach to competition is that technologies and their selection environments co-evolve.

Whenever there are economic differences there is scope for selection. But variety itself is not sufficient, the differences must be stable relative to the speed with which selection operates. In a world of perfect adaptation there would be no scope for selection. Selection is quite consistent with random variation (Monod, 1971) but what it does require is elements of inertia to hold competing varieties in a form long enough for selection to operate (Matthews, 1985; Hannan and Freeman, 1977). It is here that organizations become crucial, for one of their attributes is to create structures of thought and activity that are impervious, in part, to adaptive pressures (Hrebeniak and Joyce, 1985). Organizations generate variety and they hold variety sufficiently constant for selection to operate. As Hannan and Freeman (1977) put it, rather graphically, "Failing churches do not become retail stores nor do firms transform themselves into churches" (p. 957). Similarly, within and between competing technologies, firms generate commitments and loyalties which are not easily shaken. Adaption and selection both have a role to play, but we do insist that the ability to adapt depends in part on the past history of selective experience of the firm. Cases abound, for example, of firms where past success in the selection environment has lulled them into a false sense of security, minimizing their adaptive response when adaptation was most needed.

Of course, to interpret the evolutionary argument solely in terms of Darwinian selection would be a crude error. Technologies are articulated by purposeful organizations capable of search activity and capable of reacting, although often erroneously and within limits, to anticipated events. There are plausible arguments for claiming that the nature and timing of inventions are random events but, equally, there are powerful inducement mechanisms at work in shaping the rate and direction of inventive activity. Certainly the transition from invention to innovation is guided by selective forces. In terms of evolutionary theory, there is a clear Lamarckian element to be incorporated here. Not only do innovations arise in response to perceived needs and opportunities, they are carried through time in the memory of firms and other institutions in such a way that the experience of the past shapes what can be achieved in the future. The fact that firms learn, have memory, and possess mechanisms for maintaining memory over time in the face of changes in personnel, is the source of the chief elements of irreversibility and path dependence in the pattern of economic progress.

Finally, one must not draw too sharply the distinction between firms and their selective environment. Alliances with other firms to share the market, or to perform co-operative research are common phenomena, as are attempts to sway governments in favor of protective tariffs, production subsidies or advantageous technological standards. In doing so firms can change the selective pressures that they experience to their advantage.

However, it is our view that these qualifications enrich rather than diminish the significance of evolutionary thinking in this area. The bedrock of competition remains variety and selection, and no mechanism for generating variety is more potent in the long run than that which stimulates technological change.

Concepts and Definitions

Before continuing with the main themes of this paper it will be useful to clarify some of our concepts. The analogue to a species is a set of products and their methods of production that are drawn from the same technological knowledge base. They are the units of selection. The economic weight of a product or process is measured by its prevailing share in economic activity within a specified selection environment. It is the evolutionary process which changes the relative shares of the co-existing technologies as we shall indicate in Section IV. The market is defined as a homogeneous field within a selection environment, homogeneous in the sense that all the products and processes competing in that market are subjected to the same selective pressure. A selection environment which consists of more than one market is said to be segmented. Finally, the firm is defined as an organization articulating a knowledge base to generate a particular revealed technological performance in pursuit of certain objectives. This is not coterminous with the firm as traditionally defined in terms of control over the disposition of capital assets. Rather, in contemporary conditions, our "firm" is to be interpreted typically as a business sub-unit of a larger enterprise, the larger enterprise often being multitechnology in nature. The relations between the "firm" and the larger "umbrella enterprise" often constitute an important part of the operating environment of the former. Indeed the identification of the boundary relationships between the operating sub-unit and the umbrella is one of the more important tasks which any empirical study of innovation must undertake. This boundary is of particular significance when we come to the matter of levels of technology strategy and their interaction between the business unit and the corporate body. Note finally that our treatment of the firm does not require that it *maximize* profits or any other performance attribute. All we do require is that it seeks *improved* performance, improved in terms of certain well defined attributes.

II. TECHNOLOGICAL COMPETITION

Within the dominant schools of economic theory, competition is a state of equilibrium, based on two quite different premises; that each firm has no power to influence market prices, and that actual or threatened entry establishes a position of normal profitability for each firm. From a business perspective this view of competition is all rather puzzling as many scholars have argued (Hayek, 1948; McNulty, 1968). Indeed, Morgenstern goes so far as to claim that economists' use of the word competition has lost touch with reality, precisely because it eliminates any connotation of struggle and rivalry. Whatever the merits of the equilibrium view, and they are considerable, they are simply inappropriate to the study of technological change. Here the appropriate perspective is of struggle and rivalry, of a process of competition between unequals. Superior product and process technology is a basis for superior profitability that in turn gives the firm potential advantages in all those competition enhancing activities which require an investment of resources. Whether it be capacity expansion, marketing activity, training and skill enhancing activity, or innovation, all of these key competitive activities are resource based. Command of resources is thus a necessary, if not a sufficient condition for maintaining or enhancing a competitive position.

Thus our evolutionary view about competition is concerned with a process of change, driven by technological differences between firms that have as their outcome continuous changes in the relative economic significance of the competing technologies. The heart of firm strategy is to generate and maintain advantageous competitive differences over one's rivals (Itami, 1987). This perspective on competition raises questions at three distinct levels. The first concerns the sources of technological variety across firms. Why do firms differ in their revealed technological performance at a point in time? And, why do firms differ in their creative ability to advance their revealed performance over time? The second set of questions concerns the operations of the selection environment. How are different technologies evaluated, and how quickly are the effects of this evaluation translated into changing economic weight? The third concerns the behavior of firms. In particular, how do they translate profits into enhancement of market share, and why is it that they are not infinitely malleable in the face of competing technological advances? Why, that is, is there loyality and inertia so that technological differences can persist long enough to generate a distribution of co-existing technologies in an industry?

To provide a perspective on these questions we shall proceed in two stages: developing concepts to analyze technology; and developing a framework to illustrate the operation of a simplified competitive, economic selection mechanism.

III. THE CLASSIFICATION OF TECHNOLOGY

A. Technological Regimes and Design Configurations

In this section we outline a perspective on technological change which emphasizes two central ideas: (1) that innovation is a primary source of the variety with which selection works, and (2) that technologies define an agenda for change to be exploited by a related sequence of innovations. The competitive performance of a firm in the short run depends on the position of its technology within the relevant technology distribution. In the long run it depends on the ability of the firm to maintain a momentum of technological improvement within the constraints of the relevant agenda. Thus, we argue, competitive performance depends not simply on success at a single innovation but rather success at a sequence of innovations and related postinnovation improvements (Georghiou et al., 1986). Many firms enjoy a single innovation success but relatively few maintain the sequence of innovations which underpins a dominant long run competitive position. Hence the relevant set of technology concepts has to capture an element of technological continuity while permitting significant dimensions of technological change. Our concepts of technological regimes and related design configurations are meant to capture the elements of continuity, while the changes in revealed performance characteristics capture the elements of development (Georghiou et al., 1986, Chapter 2). This involves a shift in perspective from treating innovations as isolated, discrete events, to treating innovations in terms of an evolving flow of developments within the confines of a technological agenda.

Both the knowledge base of a technology and its revealed performance exist within a particular structure that we describe in terms of regimes and configurations. A design configuration is a particular set of facts, hypotheses, operating procedures (know-how and know-what), and design parameters that enable energy and materials to be translated into products and processes with a particular physical configuration and embodying a particular level of revealed performance. Any configuration permits a range of performance levels to be reached, each one associated with a distinct product or process, and these refinements are typically achieved via a sequence of cumulative, incremental changes. By altering some components of the knowledge base other design configurations may be created to provide alternative avenues for articulating a particular set of functional performance characteristics. The collection of design configurations, which share elements of a common knowledge base we term a technological regime. The regime defines the core knowledge base for the set of design configurations, a core which is shared by all the firms in the regime and which sets boundaries to the productive activities which can be undertaken by any of the firms involved (Richardson, 1972). In short, regimes are aggregated from the knowledge bases around which firms coalesce in specialized but related activities.

In practice, where one draws the line between regimes and configurations is somewhat arbitrary and should be guided simply by the objectives of the investigation. Some examples, drawn from our studies of technological competition will help to illustrate their meaning. Within electronics communication technology, optical fibre and co-axial cable systems define different technological regimes, with different core knowledge bases, the one based on the transmission and reception of photons, the other on the transmission and reception of electrons. Within optoelectronics communication systems we find different design configurations with combinations of monomode or multi-mode fibre and laser or light emitting diode devices. Within the UK telecommunications industry, different firms are specialized within different design configurations but all are aiming to supply users with a similar set of performance characteristics. As a second example, consider the development of cardiovascular drug technology since 1960. At that time, ICI developed a new way of improving the health of heart patients, a method which rested on the radically new idea of protecting the heart from the action of the body's own hormones, in particular, adrenalin. The sequence of β-block-

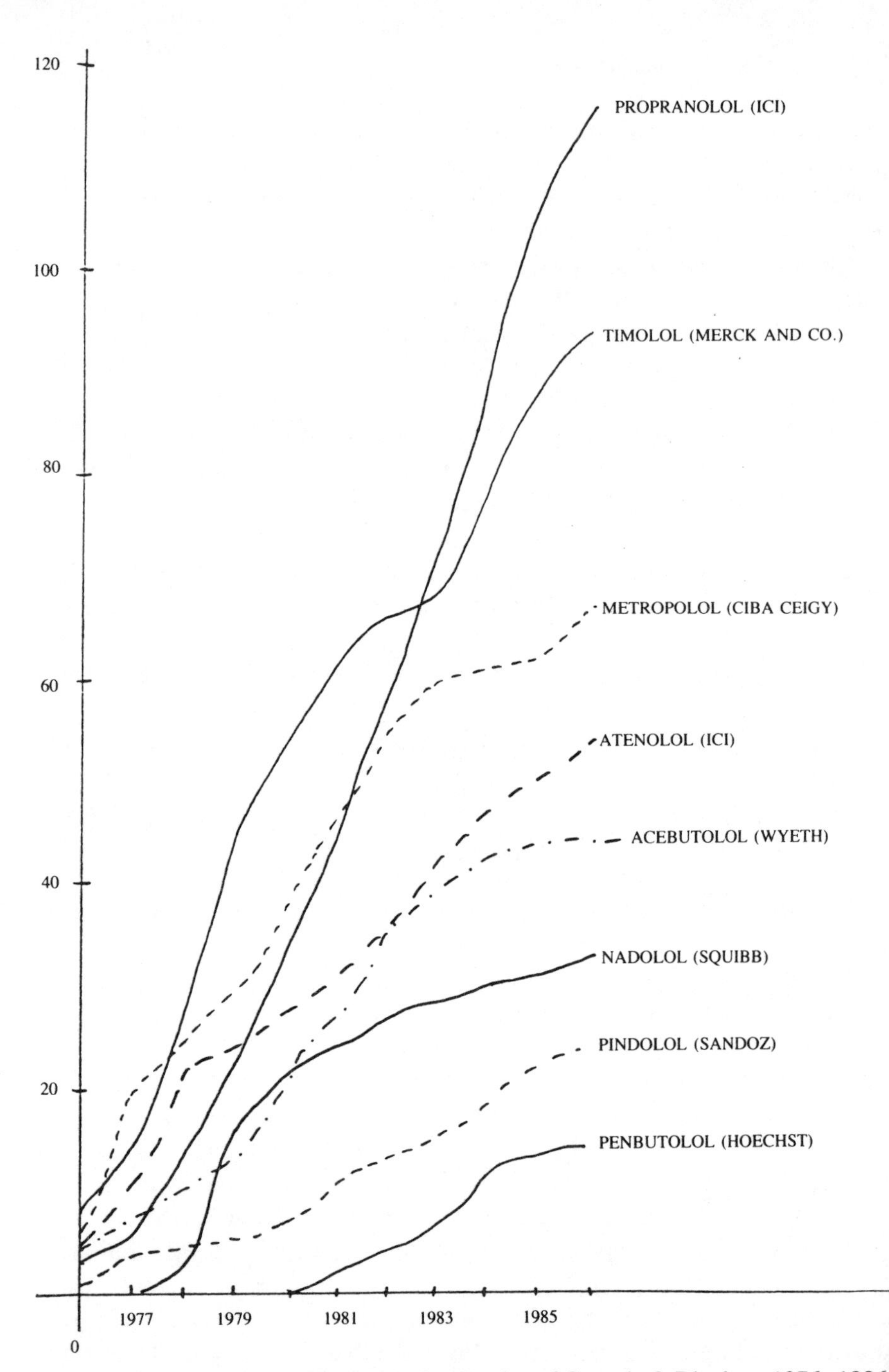

Figure 1. Cumulative World Market Entries of Generic β-Blockers 1976–1986 (and companies)

er drugs that followed created a new technological regime. Within this regime many different design configurations have developed, drugs with different chemical structures and paths to synthesis, and with different side effects. Figure 1 illustrates the pattern of world market entry for the major drugs introduced to 1986, and illustrates the variety within the regime. Within this evolving regime, ICI has been able to maintain its dominant world market position only through a sequence of improvements and new configurations: improvements which reduce the daily dose rate and combine the β-blocker with other cardiovascular drugs; and new configurations which select between the effects of different hormones on the heart's action. Tenormin, ICI's cardioselective drug is still considered to be the world leader in terms of performance characteristics despite being introduced in 1976. Market statistics show that ICI's two design configurations still accounted for 49% of worldwide β-blocker sales in 1985, twenty years after the first drug, Inderal, was launched in the U.K. Within this technology we find three levels of technological competition. Once patents expire, the generic drug producers are free to produce within any chosen design configuration, and do so in the U.K. at prices some 80% below that of the branded product. Between configurations there is also intense competition with each major pharmaceutical company, loyal to its own drug. Finally, we have seen the emergence of two new technological regimes pioneered by different companies. Pfizer and Bayer have introduced calcium channel antagonists, while Squibb has introduced ACE-inhibitors, drugs which work in quite different ways from β-blockers and each other, and involve quite different core knowledge bases. Calcium channel drugs, in particular, have proved a significant threat in the angina market and now account for 26% of world prescriptions as against 20% for β-blockers. Finally, as is so often the case, the traditional regimes, based upon nitrates, have also been improved over time, to produce intense competition between four main technological regimes championed by different companies.

The purpose of these concepts, regimes and configurations, is precisely to help structure the relationship between variety and competition and also to explain one source of organizational inertia. It is our hypothesis that the structure of the knowledge base differs sharply between regimes and to a lesser extent between configurations, and that differences in structure are associated with different mechanisms of technological advance. For this reason alone, the attempt by any firm to change its knowledge base, whether by internal effort, imitation, or technology transfer arrangement, is likely to prove extremely difficult across regimes but much easier across related design configurations. Regimes differ according to the proportion of knowledge which is discovered by scientific or empirical means; they differ in the division of knowledge between codifiable, publicly available, and tacit firm, specific forms; and, they differ according to their dependence on other knowledge bases that are generated outside the industry. Thus the way knowledge advances and the ease of its appropriability can differ even between configurations in the same regime (De Vincenti, 1984). Once a firm has acquired a structured knowledge base it becomes with time easier to advance within it than to change to a new configuration. Specialization creates its own barriers to future change.

Within this scheme of ideas, the distinction between radical and incremental change falls naturally into place. Evolutionary, incremental change takes place within design configurations, it is often driven by interaction with suppliers and users (Rosenberg, 1982) and it is naturally limited by the agenda of the configuration. This pattern of change generates the classic 'S' curve of technology development (Jantsch, 1967; Foster, 1986) with the limit being determined by the onset of sharply diminishing returns to investments in improving technology. Radical innovations, are discontinuities which establish new configurations or new technological regimes (Tushman and Anderson, 1985). While improvements within configurations arise within a particular population of firms, radical innovations have a high probability of being introduced by outsiders (Gilfillan, 1935). These same concepts also clarify the origins of retardation and acceleration in technological progress. For any given design configuration, limits to its agenda must ultimately enforce the onset of retardation. Progress within the regime must then be contingent on discovering new and superior design configurations but even this prospect will be limited. Ultimately, the maintenance of progress is contingent upon the introduction of new technological regimes. One may note here the parallel with the distinction between gradualism and punctuated change which have developed within evolutionary biology (Gould, 1980). Finally, we emphasize how these concepts lead to different levels of competition, between regimes and configurations within a regime, and within a given configuration. The different forms of competition differ crucially in the mechanisms by which firms differentiate themselves and advance their revealed technological performance. The potential for competitive differentiation is clearly least within design configurations, where there is a greater commonality of knowledge base, and greatest between regimes where firms may have few elements of knowledge in common.

B. Revealed Performance and the Technology Set

Associated with the design configurations within a regime are sets of products and methods of production to which we assign functional performance characteristics. Users buy a product because of the characteristics it contains and the producers employ a production method because of the inputs which it employs. The technology jointly determines product and process attributes, and the scope for changing the product without concommittant changes in the process is often narrowly circumscribed (Abernathy, 1978). Over time one can then map the evolution of the technology in terms of the rate and direction of change in these performance characteristics (Sahal, 1981; Saviotti, 1984).

However, the basis for an evolutionary approach to technological change lies not in terms of performance characteristics alone but in their conjunction with the economic evaluation of these characteristics in the market place. For each product one may assign a unit cost and a price, based upon the prices of productive inputs and the market evaluations of its product characteristics. Different market environments may then generate quite different price and cost rankings of the different technologies. To fix ideas, consider a homogeneous market environment which evaluates each technology in identical fashion with input price vector w and product characteristics valuations vector v. If a is the vector of inputs per unit of output, and α is the vector of characteristics per unit of product then we can define $h = wa$ as unit cost and $p^* = \alpha v$ as unit value (quality adjusted) of a given product. Repeating this exercise for all competing products in the market at a point in time allows us to draw the selection set shown in Figure 2a. It is within this set that the competitive process operates.

As drawn, the main products are grouped within three different design configurations A, B and C. Each point represents a particular product with a given level of unit cost, h_i, and an associated quality adjusted price, p_i^*. The set may be densely populated by competing products or it may be reduced to only two alternatives, for this is the minimum number consistent with competition as a process. The set represents the amount of economic variety within the population of competing technologies, and its morphology is contingent upon revealed technological performance within each regime and the manner in which the environment evaluates that revealed performance. Over time, two forces can reshape this selection set. Changes in input prices and change in characteristics valuations, which by changing h_i and p_i^* respectively will re-order the boundaries of

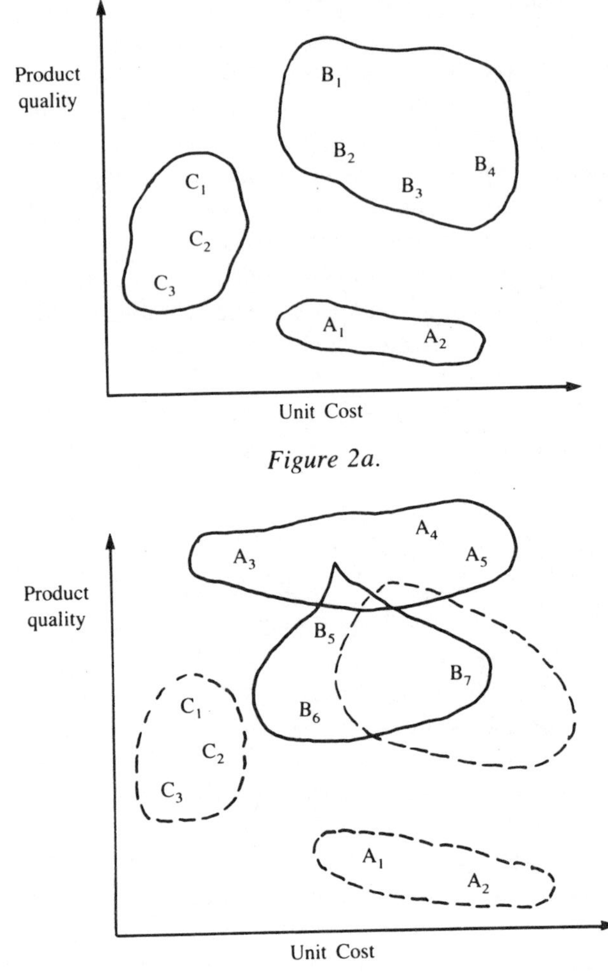

Figure 2a.

Figure 2b.

the set and change the distance between the competing products and their processes. More potently, perhaps, the addition of new regimes and improvements within regimes will also redraw and re-order the selection set, mapping competing trajectories of technological improvement in terms of product quality and unit cost. The fact that configuration A appears inferior to B and C (in Figure 2a) does not rule out the possibility that at some future date it may become superior (Figure 2b). At a point in time the selection set provides a snapshot picture of economic variety.

Over time it becomes a device for charting the history of the development of technologies. It also provides a basic tool for analysing technological competition. In short, it represents the raw material upon which competitive selection can operate.

C. Knowledge Base and Organization

We come now to the second axis along which any technology can be analysed, the knowledge base, and its links with the organization as a repository of knowledge (Winter, 1982). We shall suggest that variety in revealed performance is inseparable from variety in the ways in which organizations structure and articulate their knowledge bases. Although information may exist in data banks, knowledge (i.e., structured information) can only exist in the mind of individuals. The effective knowledge base of the firm depends then on the way in which it coordinates and pools the knowledge of its constituent members. Some coordination is necessary because specialization and division of labor implies that each individual has command of only a small part of the relevant knowledge base, and while this can generate efficiency and learning economies it may also limit individual mental flexibility. Furthermore, since many decisions, and especially strategic decisions, are made in circumstances of partial ignorance (Loasby, 1976), specialization becomes a way of coping with the bounds of rationality that arise from the naturally limited cognitive and computational powers of individuals.

But the need for coordination creates a structure to communicate, filter and pool knowledge, a structure which is a significant capital asset for any firm. Within this structure, there is associated a framework for thought and for distinguishing significant from insignificant events; a framework which gives the organization a world view. Moreover, since knowledge and theories are always imperfectly specified there is room within the framework for conflict within the organization over the interpretation of events and their significance, and hence for variety in policy, strategy and behavior.

In this way we can view the organizational structure of the firm as an operator, translating individual knowledge into collective, shared knowledge, within a chain of communication and decision making. The nature of this operator is to determine who "talks" to whom about what and with what authority. From this we may see immediately why firms come to have different knowledge bases and possess differential abilities to change their knowledge base. Different individuals and different modes of organization lead directly to different interpretations of the world within and between organizations. Some familiar empirical themes emerge here, the significance of different links between R&D, marketing and production activities, the role of product champions and entrepreneurs, and the appropriate organization of the R&D function, be it on centralized or distributed lines. Given individual specialization and the interrelated nature of the firms' knowledge structure it is easy to foresee the difficulties that are often associated with technology transfer, joint ventures and the "not-invented-here" phenomena. External knowledge is in effect an injection of a "foreign body" within an existing specialized and interconnected structure. Not surprisingly, a firm can soon produce its own antibodies!

From these considerations we are led directly to the view that the knowledge base of a business unit coalesces around a design configuration, and that the organizational structure builds within it a growing commitment to this design configuration, both limiting and shaping how the organization reacts to external technological developments. With experience and structure comes commitment and inertia so that the mind set of the firm becomes "canalized" and permits only of certain "internally consistent" paths of future development. As Dosi (1983) has persuasively suggested, the design configuration becomes a paradigm from within which to explore a self-limited world. If a firm finds it difficult to "think" differently it will equally find it difficult to "act" differently. Thus, the knowledge base and its supporting organizational structure forms a powerful mechanism to explain why firms systematically and seriously underestimate the significance of new design configurations and regimes (Cooper and Schendal, 1976). The organization becomes a powerful mechanism for ensuring the persistence of technological variety across firms. In short, it does not easily abandon that which it has successfully acquired, and finds it possible to entertain development only within the boundaries associated with its own configuration. The firm sinks investments not only in physical capital (Frankel, 1955) but in its organization and its knowledge base and from this inheritance of the past come the chief sources of delayed adaptation to changed circumstances.

Even if a number of firms within a regime could be considered to start from a common knowledge base, the certainty of their differential creativity will soon demolish this state. Based on their different personnel and organization structures, combined with random creativity and different abilities to manage research and learning processes one soon finds significant differences in the intellectual capital of the different firms.

What can be thought in the future becomes severely circumscribed by what has been thought in the past. Hence change is often cumulative and incremental within the configuration. Note that this situation denies neither the possibility of internal step changes in the firm's knowledge base nor the possibility of its adapting to outside technological stimuli. It suggests only that such possibilities become less likely, the longer it is that the firm has exploited its particular design configuration, and the more successfully it has done so. In this way it is quite plausible to argue that the technology of the business unit becomes progressively less fluid over time (Abernathy, 1978). Its parent, corporate body may enjoy greater flexibility, but not even here without limits on what it can manage in an efficient manner.

Not only is the knowledge base underpinning a regime structured within the various firms it is also partitioned between a number of different institutions, including universities and other publicly funded research establishments, not necessarily in the same country. How a firm interacts with the other institutions, each of which may have a comparative advantage in generating different parts of the knowledge base and be responsible for different advances at different times, is a factor of considerable importance in determining its creativity. While much of this knowledge may be considered in the public domain it is not thereby accessible to all on equal terms. The anterior knowledge base of the firm and its ability to manage and integrate gatekeeper activities within its operations will be important determining factors in its ability to exploit exterior knowledge (Teece, 1987). The extent of the knowledge base is also closely related to the determination of the boundaries and degree of specialization of the firm. A particular knowledge base often offers complementary avenues for integration in some directions, but effectively forecloses them in others (Richardson, 1972). Thus the degree of vertical integration is not only a question of scale economies associated with physical capital but rather hinges additionally on the costs of acquiring supplementary knowledge and exploiting the subsequent knowledge capital over a sufficient scale of output. Contractual issues of the kind stressed by Williamson (1986) also become relevant. For knowledge and changes in knowledge are by nature imperfectly defined, and when it becomes important to determine the specifications of complex products and processes on a frequent basis there may be no effective alternative to the vertical integration of the enterprise across different knowledge bases.

While it is the knowledge base which underpins revealed performance, the link between the two is by no means straightforward. Organizational structure mediates knowledge and revealed performance both in terms of the ability of the firm to design out of its knowledge base, and in terms of its ability to achieve particular levels of process control and product quality. Technology and organization are not independently determined at the level of the business unit, they are mutually determining. For these reasons alone it is impossible to understand differences in revealed performance without admitting differences in design capability and operational organization. Here again we find potent sources of technological variety.

Corresponding to the three different forms of technological competition previously outlined we find different sources of the revealed technological performance of firms. Differences may arise from the location of the firms in different regimes, or in different configurations within a regime, or from their respective design and operational capabilities which differentiate firms within the same configuration. There is clearly no implication that even if two firms possess the same knowledge base that they will reveal the same technological performance. Indeed in any organization it is possible that significant elements of its knowledge base may be underutilized or misdirected (Penrose, 1959).

Although there can be little argument that the knowledge possessed by a firm is crucial to its success and survival, the idea of a knowledge base is rather abstract and difficult to identify in practice. It may help to attempt a more formal statement of the argument by distinguishing the "elements" in the knowledge base from the "skill levels" with which they are associated. Indeed we have already taken the first step toward this distinction, when we suggested that the relationship between the knowledge base and the firm's revealed technological performance is subject to considerable variation. Following the work of McElvey and Aldrich we may relate the knowledge base to the concepts of a dominant competence and compools. We suppose that the knowledge base consists of a number of distinct elements, each element providing a portion of knowledge necessary to achieve the transformation of materials and energy into the product. Which materials to use, how to process them, in what order to carry out the stages of production, and how to put the various components together into a product would be typical examples. Within the organization each element exists as a particular level of skill embodied in individuals and teams. Collectively these elements and levels of skill define the dominant competence which allows revealed technological performance to be articulated. Firms may then differ in three ways. They may start with different elements; they may operate them at different levels of skill; and, they may structure these skills differently into a dominant competence. It is the differential skill element that makes it possible to identify, conceptually or empirically, the distinctive elements in a firm's dominant competence. Those elements, that is, which confer upon it particular advantages or disadvantages in the competitive struggle.

Figure 3 depicts a technological regime consisting of three different design configurations. A involves elements r m p z t w h b c, B involves elements r m p z w q u e s, and C involves elements r m p z t a c l. From this we see that the core of the regime involves the elements r m p z which are contained in each design configuration, and which define the intersection of the three configurations. The individual configurations are differentiated by adding elements, which are missing in one or both of the other configurations. Although the skill levels in any configuration may vary and the elements be structured in different ways, the elements themselves are invariant within the configuration. They provide the ongoing stability of the design configuration, while changes in skills and structure underpin changes in revealed performance within and between different design configurations.

The union of these sets of dominant competences, the compool (McElvey, 1982), defines the regime as a whole. Within this aggregate compool, the core elements are likely to constitute publicly available knowledge, while the remaining elements are more likely to be tacit and firm specific. The fact that the design configuration relies upon a particular organization of elements and skills is what gives the knowledge base its structured form and helps explain the differential permeability of dominant competencies to external developments in science and technology.

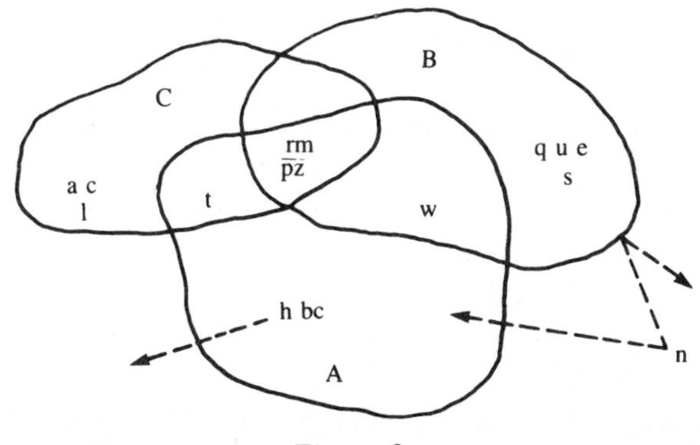

Figure 3.

When element "n" is developed it is rejected by, or is incompatible with, configuration B but it is incorporated in A at the expense of element h, so creating a new design configuration A^1. It is our hypothesis, that where "n" is absorbed into A, this event can be identified by the appearance of a new skill within the set which defines the dominant competence for that regime.

We have already suggested that the organizational structure of the firm may be treated as an operator which combines individual levels of knowledge into a collective knowledge base. More precisely, we can now suggest that the operator combines the elements of knowledge and the skills of individuals to form the dominant competence which enables performance to be revealed. The key aspect of a dominant competence is the fact that it is an organized stock of knowledge and skills. As organized, it may be linked to set of blue prints for converting certain raw materials into products and processes. But while the blue print need not be fixed once and for all, there is a relationship between the blueprint and the organization of the stock that precludes very much in the way of radical experimentation. Thus, for many firms, the initial organization of the dominant competence sets the agenda for future development, and firms differ in revealed technological performance because each has organized its dominant competence in a slightly different way and so has launched itself on its own trajectory of development. The firms which will come to dominate in a particular sector of the market in the long run, we hypothesize, will be those whose organization is most flexible in the sense that their flexibility allows for a continuous stream of adjustments to the fundamental blueprint as competitive and environmental pressures demand. It follows from the above that regimes and configurations are devices for structuring and limiting the differences in revealed performance between business units both at a point in time and in the way they develop over time. Differences in revealed performance should be at a minimum, although not zero, within regimes and at their greatest between regimes. The mechanisms which determine the differences and similarities in knowledge bases between business units is a matter for detailed empirical investigation. Internal knowledge generating activities combined with secrecy play an obvious role. Their operation has to be judged against the nature of the knowledge they produce and against the existence of informal and formal mechanisms for sharing knowledge between firms, mechanisms that range from patent citations to networks of professional contacts, through the job mobility of personal and on to collaborative ventures in technology development.

In summary we have argued that the technological element in competitive performance needs to be looked at in two dimensions. Firms reveal technological performance by articulating an underpinning knowledge base. Due to factors partly concerned with the nature of knowledge e.g., whether it is codifiable or tacit, and partly due to questions of skill and organizational structure we expect there to be variety in revealed performance and in knowledge bases. The forces which are at work make for cumulative differences, and ensure that once differences are established they are not easily eliminated.

So far our argument has been entirely focused on the origins and persistence of technological variety. The question that this raises for us is what sense are we to make of variety in the context of competition. What difference do these distinctions make to the understanding of technological development and economic change? It is here that the evolutionary perspective with its focus on distributions of phenomena comes to the fore. To the elaboration of this we now turn.

IV. TECHNOLOGICAL VARIETY AND THE EVOLUTIONARY MECHANISM

The first aspect to be clear about in any evolutionary theory of technological competition is the shift in perspective from matters of equilibrium to matters of change, from the scale of production of competing commodities to the rates of growth of those scales of production. The second aspect to emphasize is that selection operates with respect to the performance characteristics of the competing products and processes, and only indirectly with respect to firms. Products and their processes are selected jointly and directly, according to their overall performance. A single product firm is naturally subject to the same selective process as is its technology, while the selective process on a multiproduct organization depends on the balance of selective forces across its product and process portfolio. This separation does not rule out the possibility that the business unit from which a product emerges will make a difference to the market's evaluation of that product.

At this point we employ the distinction, which is crucial to much organizational theory, between the firm, i.e., business unit, and its operating environment. This distinction has proved extremely fruitful in the analysis of markets and hierarchies in industrial organization (Williamson, 1986; Teece, 1987), and in the contingency theories which relate "good" organizational structures to environmental attributes (Terreberry, 1968; Lawrence and Lorsch, 1967; Burns and Stalker, 1961). It is equally invaluable to the analysis of evolutionary processes. Needless to add, drawing this boundary is often problematic, especially where account is taken of links with an umbrella organization and of the actions which firms may take to change their environment. Moreover, one must certainly distinguish technological from market environments since firms may collaborate in the one while competing in the other. Nonetheless, the distinction between firm and environment remains central to evolutionary theories of competition.

Within this perspective we may assign to the firm three attributes. First, there is the efficiency of the firm as embodied in its revealed technological performance, the productivity with which it transforms inputs into products and the functional characteristics of those products. As we argued in the previous section, revealed performance depends on the blend of competences and skills structured within the organization, which in turn are located within a bounded design configuration. Second, there is the propensity of the firm to grow as measured by the relation between its growth rate and its profit margin. Growth depends on access to internal and external finance, on the investment requirements to expand capacity and marketing activity, on the ability to manage growth without sacrificing efficiency, and on the simple willingness to grow. It will also depend on the strategic investment and financial policies of any umbrella organization. In previous work (Metcalfe and Gibbons, 1986) we have summarized these factors by the term fitness, since, despite possible ambiguities, it captures the essential link between efficiency and growth. Finally, there is the creativity of the firm, the ability to improve revealed performance through knowledge base enhancing activities; learning phenomena, formal R&D and superior design capabilities. Creativity depends on the resources available to advance technology; the opportunities for advance within the chosen configuration; the incentives to advance in relation to scale of effort, the scale of application of advances and their appropriability; and, the effectiveness with which the firm and the umbrella organization manages its creative activities.

We expect that any two firms will differ in all three dimensions. Variety in creativity naturally leads to variety in efficiency, while variety in 'fitness' leads to different resource bases for financing subsequent technological advance. Following our previous discussion we also expect that these differences will not be easily eliminated. Inertia sustains variety in all three dimensions.

On the environmental side we have a number of difficulties to face. While several scholars have attempted to crate environmental typologies (Terreberry, 1968; Jurkovitch, 1974; McElvey, 1982) they must be considered as being of limited success. Attributes such as turbulence, complexity and uncertainty are undoubtedly relevant but without further specification they admit of no clear interpretation. In part this is because it is perceptions of these attributes which is important, and in part it is because the theoretical link between attributes and behavior is not well specified. Within an evolutionary approach we can be more precise. In particular, we can specify the market environment in terms of the following considerations.

First, the market environment generates a set of input prices and product characteristics valuations that translate technological variety into economic variety. This provides the basis for constructing the selection set. However, while the input prices are normally explicit market data, the product performance valuations are almost always *implicit* to be discovered by the firm through market research activities and interaction with its customers (Rosenbloom and Abernathy, 1982). In short, the environment generates a price structure which may, or may not, be the same for all firms, and certainly may be perceived differently by different firms. The second attribute of the environment is the rate of growth of the market, which is typically subject to retardation over time. The third attribute is the degree of selective pressure which the market imposes. This depends on the frequency with which selective decisions are made, frequent selection generating fine grained environments and occasional selection generating coarse grained environments (Levins, 1968). Fine grained environments may be created through frequent repeat purchases or by a steady flow of different customers. It depends upon the severity with which selection operates, how quickly firms are punished or rewarded by their deviations from average behavior. Finally, it depends upon the uniformity of selection, a uniform market being one in which all firms experience the same selective pressure, a segmented one being one in which pressure varies across firms, e.g., because of goodwill or long-term contractual relationships. In an obvious sense, uniformity of selection requires informed, intelligent customers. The fourth attribute of the environment relates to the manner in which it changes over time. A tranquil environment is defined as one with a given structure of product and process characteristic valuations, which grows at a constant compound rate. Degrees of turbulence may then be defined relative to various shocks to the growth rate and price structure, some of them being exogenous and some of them being endogenous to the selection process.

A. Selection and the Variety of Technology

We can now use these concepts to show how selection works on the technology set to change continually the relative economic weight and the survival prospects of competing products and processes. In effect, this creates a multitechnology diffusion process tending to converge upon a dominant design.

In order to emphasize the central ideas, we consider a highly simplified selection process across a given set of technologies with a variety of products derived from different design configurations (cf. Figure 2a). The extent of technological variety is given for analytical purposes. The market environment has a given price structure, independent of the selection process, and grows at the compound rate g_d. The selection environment is uniform and operates continuously. On the basis of the given price structure we can identify any particular product and its associated process with a unit cost of production, h_i, and a quality adjusted price, p_i^*. If w_j and v_k are the prices of the jth input and kth product characteristic respectively, then we can write $h_i = \Sigma w_j a_{ij}$ and $p_i^* = \Sigma \alpha_{ik} v_k$, with a_{ij} the input of j into a unit of i, and α_{ik} the output of characteristic k per unit of i. It will help to simplify the expression for p_i^* by choosing one of the v_k as an index of the entire price structure and writing $p_i^* = v_o \alpha_i$, α_i representing the "quality" of the product in terms of the index characteristic. The actual price of a product we represent by p_i, and, in general, this will differ from p_i^*. On the basis of this information we then have the selection set in Figure 4. Note that the choice of a different index characteristic, or any change in the price structure will require this set to be redrawn. All the competing products lie on or within the boundaries of this set but they need not be evenly distributed with it and regions of the set may be empty.

The selection mechanism that we now analyze is remarkably simple in its structure. The environment translates technological variety into economic variety and determines the economic distance between the competing technologies. The capacity to produce any one commodity increases in absolute terms if its production is profitable. But it only increases its relative significance if it is more profitable to produce than the average for the technology set. The further above (below) average practice is the product the faster the rate of increase (decline) of its market share. However, selection, by changing the relative weights of the competing technologies, continually redefines the levels of average practice performance, pushing some products into bankruptcy and changing the distance

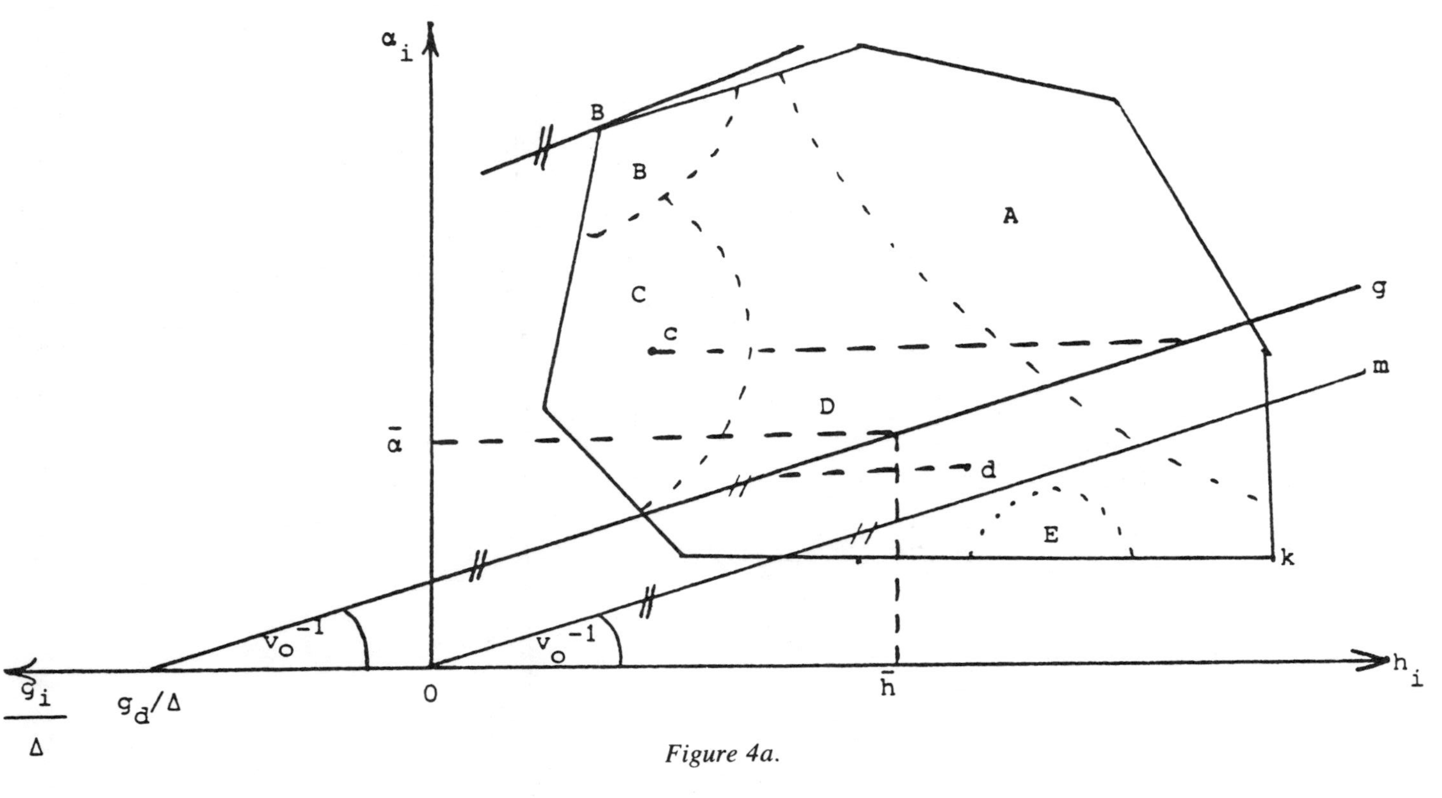
α_i
B
B
A
C
c
g
m
D
$\bar{\alpha}$
d
E
k
v_o^{-1}
v_o^{-1}
$\frac{g_i}{\Delta}$
g_d/Δ
0
$\bar{h}$
h_i

Figure 4a.

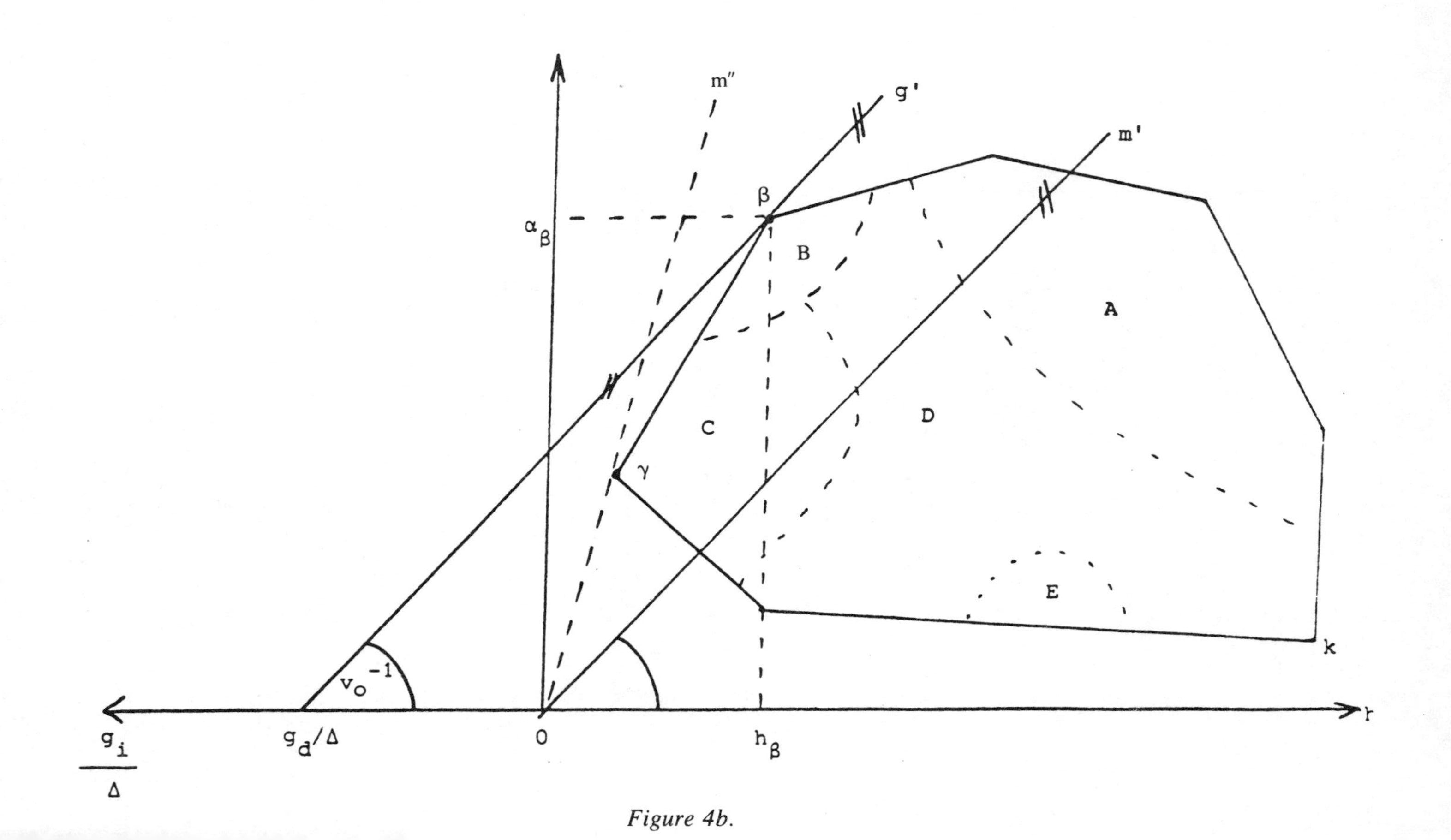

Figure 4b.

of each surviving product from average practice. The rate of change of average practice depends directly on measures of the variety contained in the technology set. In short, the selection mechanism is profit driven, because profits provide the resources with which a firm changes the economic significance of its product. In turn growth depends upon the dynamics of capital accumulation and the process of customer selection. To explore these ideas in more depth we must define various statistics which capture the variety contained in the selection set, and here we need to specify the economic weights of the different technologies. These weights we define by the market shares of each product in the total value of output produced by the firms in the technology set. Indicate the shares by s_i. Then we can define, average practice unit cost, $\bar{h}$, average product quality $\bar{\alpha}$, average quality adjusted price $\bar{p}^*$ and average market price, $\bar{p}$, as follows,

$$\bar{h} = \Sigma s_i h_i, \; \bar{\alpha} = \Sigma s_i \alpha_i, \; \bar{p}^* = v_o \bar{\alpha}, \text{ and } \bar{p} = \Sigma s_i p_i$$

Similarly we have the corresponding measures of variance and covariance defined by,

$$V(h) = \Sigma s_i (h_i - h)^2, \; V(\alpha) = \Sigma s_i (\alpha_i - \bar{\alpha})^2, \; V(p^*) = v_o^2 V(\alpha)$$

$$V(p) = \Sigma s_i (p_i - \bar{p})^2, \text{ and } C(h,\alpha) = \Sigma s_i (h_i - h)(\alpha_i - \bar{\alpha}).$$

These statistics provide the fundamental data to measure the rate and direction of evolutionary change across the given technology set.

Within the competitive process, evolution is guided by three rules; the rule of survival, the rule of accumulation, and the rule of customer selection. Survival of a product requires that it earn nonnegative profits, $p_i - h_i > 0$, and correspondingly the survival of single product firms is equivalent to survival of their product. Bankruptcy entails the elimination of the product or process from the technology set. The rule of accumulation specifies that firms making above normal profits ($p_i > h_i$) have the potential to grow, and do grow at a rate which depends upon their propensity to accumulate. We let each firm concentrate its capital investment entirely on its own product, even where it is not the best-practice technology, by reinvesting internally generated profits and supplementing this with external finance. The higher the firm's profit rate the greater its access to investible funds (Turner and Soete, 1984; Downie, 1955). If f_i represents the given propensity to accumulate, then the growth rate of the output and capacity of the ith product is given by

$$g_i = f_i(p_i - h_i) \quad (1)$$

where, f_i varies inversely with the investment/output ratio and positively with the ratio of internal and external funds to the capital stock (Metcalfe and Gibbons, 1986).

The rule of customer selection forms the demand side of the selection mechanism and is concerned with the factors which change the customer basis of the competing products. In this simple, unsegmented market environment each of the competing products supplies the same set of functional characteristics to customers and they are therefore perfect substitutes if they sell at their quality adjusted prices. If this were not the case, the market would have to be divided into its appropriate segments. However, this equilibrium condition need not hold continuously. During the selection process there is no requirement that each product sells at its quality adjusted price, i.e., that $p_i = p_i^*$. Any increase in the price of i, or fall in the price of j should be associated with some erosion of the customer base for product i. Similarly, an improvement in the quality of i (either it contains more characteristics, or some of these particular characteristics are more highly valued) or a reduction in the quality of j should work to enhance the customer base for product i. One way to represent the dynamics of customer selection, is to let customers switch between products in proportion to their relative market shares and the price advantages which the products offer (Phelps and Winter, 1970). Our representation of this gives a relation of the form

$$\frac{ds_i}{dt} = \Sigma_j\, s_i s_j \delta_{ij}\, [(p_j - p_j^*) - (p_i - p_i^*)],$$

where the δ_{ij} coefficients represent the probability of a demand switch upon "contact" between customers of firms i and j. Now in a uniform market environment, by definition, all the δ_{ij} coefficients take the same value, δ, from whence we obtain the following selection rule (taking account also of the definitions of p_i^* and $\bar{p}_i^*$),

$$\frac{1}{s_i}\frac{ds_i}{dt} = g_i - g_d = \delta[(\bar{p} - p_i) - v_o(\bar{\alpha} - \alpha_i)] \qquad (2)$$

A product's share in market demand is increased if it has a below average market price and above average product quality, or some favourable combination of the two to make the bracketed expression in eq. 2 positive. Notice that eq. 2 represents only one possible rule for customer selection. However, it does capture plausible attributes of the selection process, and it does satisfy one necessary logical condition, namely that

$\Sigma s_i g_i$ equals the growth rate of the aggregate market. A rule which does not meet this requirement is simply internally inconsistent. The clear advantage to be gained from eq. 2 is that it relates the dynamics of selection to the distance of the particular product from the industry average. In this way it rests upon a very simple statistic of inter-product variety.

Two special cases come immediately to mind when considering this customer selection rule. When $\delta = \infty$, we have the analogue to perfect competition. A firm must set $p_i = p_i^*$ (and hence $\bar{p} = \bar{p}^*$), for if $p_i > p_i^*$ it immediately loses its entire customer base. Conversely, with $\delta = 0$, we have the analogue to monopoly, with the firm's customer base and market share quite independent of its position in the price and quality structure. In this case customer loyalty is absolute. In between are all the cases of dynamic imperfect competition, in which the rate of change of the firm's customer base does depend on how it prices relative to the average for the technology set. It should be clear that, under imperfect competition, the firm does not have a given demand curve but rather a demand curve which shifts over time with changes in its customer base in accordance with eq. 2.

In drawing the two selection rules together we have a number of alternatives only one of which will be explored here. To focus our discussion, we consider the implications of a uniform propensity to accumulate, $f_i = f$, and concentrate solely on those situations of dynamic market balance where the growth rate of a firm's capacity equals the growth rate of its market. These balanced paths are illustrative of the secular trends of the evolutionary process. They are the basis from which we can consider other pricing policies that necessarily entail that the growth of demand and the growth of capacity for a commodity will be unequal. Since growing underutilization of capacity and increasing order books are unlikely to continue unchanged in any firm aware of its business opportunities, such alternative pricing policies may rightly be called 'short-term.'

Together, these assumptions imply that the market shares of the different products evolve according to the following condition, obtained by combining eqs. 1 and 2,

$$\frac{1}{s_i}\frac{ds_i}{dt} = \frac{\delta f}{f + \delta}\,[(\bar{h} - h_i) - v_o(\bar{\alpha} - \alpha_i)], \tag{3}$$

$$= \Delta . S_i$$

and, $g_i > 0$ if $p_i > h_i$.

Equation 3 is the fundamental equation of the evolutionary process. It indicates that the selective pressure acting on a technology is the product of the coefficient of selection and the selective force. In this expression, Δ represents the coefficient of selection, and it is increasing in both *f* and δ. The expression in square brackets, S_i, we term the selective force acting on i. Providing a product generates profits its output increases absolutely. But in a growing market this does not mean it grows in relative market significance. To increase in relative significance, a product must generate above average profitability either by having unit costs below average ($h_i < \bar{h}$) or by having above average product quality ($\alpha_i > \bar{\alpha}$) or some appropriate combination of the two. Hence the relative economic weight of a product changes according to its distance from average performance within the technology set.

To explore these ideas further, we shall concentrate on showing how the evolution of a product's market share depends on its position in the technology set. This entails a two-fold partitioning of the selection set in Figure 4a. The first step is to partition the selection set into the competing design configurations along the lines of Section III above. Five such configurations, A to E, are shown and, to simplify the diagram, they form non-intersecting sets of products and their associated processes. Along the horizontal axis of Figure 4a are measured unit costs, h_i, while on the vertical axis are measured the product qualities (in terms of the index characteristic), α_i. The second stage in the partitioning is in terms of the selective pressure acting on any given point in the selection set. This economic partitioning depends on the price of the representative, index characteristic and the given growth rate of the market, g_d. The implicit market price of the index characteristic is measured by the slope v_o. The two lines 'm' and 'g' each have slope v_o^{-1} and intercepts at the origin and point $g_d\Delta^{-1}$ respectively. They partition the technology set into three distinct areas according to the selective force at work. Below 'm' are all the bankrupt products, products for which $p_i < h_i$. Above 'm' but below 'g' are profitable products but with less than average profitability, so that they are declining in relative importance, $g_i < g_d$. Above 'g' are products with above average profitability, which correspondingly increase their market shares over time, $g_i > g_d$. Products located on 'm' just break even, while any lying on 'g' are dynamically representative with constant market shares. Now the bracketed term in eq. 3, measuring the selective force operating on a technology, has a simple interpretation in Figure 4a. Take any profitable product point (c or d), then the length of the horizontal line from this point to the average performance line g, measures the magnitude and direction of the selective force. For product d this is negative, while for product c it is positive. Thus the evolutionary principle that rate of selection equals the selective pressure multiplied by the selective force has a ready interpretation in this diagram. At a glance we can see how the market position of rival technologies will evolve. Design configuration E is out of the race, B and C both generate products of above average profitability, while A and D are configurations with mixed fortunes. Holding v_o constant we can see immediately which of the design configurations will rise to market dominance at this particular price. This will be configuration β, represented by product/process β. This follows immediately we recognize that product/process β supports the fastest balanced growth rate of capacity and demand in the selection set.

Along these balanced paths, the firms are setting actual prices which maintain long-run capacity utilization. For any firm we find that, its price is a weighted average of its quality adjusted price and its unit cost level,

$$p_i = \frac{\delta}{f + \delta} p_i^* + \frac{f}{f + \delta} h_i \tag{4}$$

and that,

$$\bar{p} = \frac{g_d}{f} + \bar{h} \tag{5}$$

so that average profitability just suffices to support aggregate output growth at the market rate.

It is a characteristic of an imperfect customer selection process that firms charge different prices for their products, differences over and above those explained by differences in product quality. In fact, $p_i = p_i^*$ only for those products on the verge of bankruptcy (on 'm'), while for all other profitable products we have $p_i < p_i^*$.

Correspondingly we find the variance of market prices across the profitable products is given by

$$V(p) = \left(\frac{f}{f + \delta}\right)^2 V(h) + \left(\frac{\delta}{f + \delta}\right)^2 V(p^*) + \frac{2f\delta}{(f + \delta)^2} v_o\, C(h,\alpha) \tag{6}$$

Special cases then emerge in an obvious way. With $\delta = \infty$ 'perfect competition', we find that $V(p) = V(p^*)$ and unit costs have no effect on the dispersion of market prices. With $\delta = 0$, a world of independent monopolies, we find that $V(p) = V(h)$ and product quality has no effect on the dispersion of prices. In between are all the cases associated with dynamic imperfect competition, for which the covariance between unit costs and product quality becomes significant.

Of course, Figure 4a can only represent a snapshot of the competitive process. As market shares change, so economic weight is redistributed within the technology set to redefine the statistical moments of the selection process. One way to measure and summarize these evolutionary trends, is to focus on the rates of change of average unit cost and average product quality, which are governed by the relations,

$$\frac{d\bar{h}}{dt} = \Delta\,[v_o\, C(h, \alpha) - V(h)] \tag{7}$$

and,

$$\frac{d\bar{\alpha}}{dt} = \Delta\,[v_o\, V(\alpha) - C(h, \alpha)] \tag{8}$$

Here we see a distinguishing feature of the evolutionary perspective, with the rates of change of average population characteristics being related to measures of variety within the population. In the simplest special case, where all products are of the same quality, $V(\alpha) = V(p^*) = 0$, we have the equivalent of the famous 'Fisher Law,' that average practice unit cost declines at a rate proportional to the variance of unit costs within the technology set (Nelson and Winter, 1983, p. 245). More generally we see that if unit cost and product quality are negatively correlated, then selection always reduces average unit cost and increases average product quality. Furthermore, in a world of monopoly relationships; $\delta = 0$, it follows immediately that evolution in our sense is not possible.

On combining eqs. 7 and 8 we then have the following aggregate measure of the rate of selective change,

$$\begin{aligned} v_o \frac{d\bar{\alpha}}{dt} - \frac{d\bar{h}}{dt} &= \Delta\,[V(h) + V(p^*) - 2v_o\, C(h\ \alpha)] \\ &= \Delta^{-1}\, V(g) \end{aligned} \tag{9}$$

where $V(g)$ is the variance in growth rates within the profitable areas of the technology set (above 'm'). This expression can also be interpreted as measuring the rate of change of average profitability in the industry, that is, the change in average profitability that would arise if all products sold at their quality adjusted prices. Since $V(g)$ is nonnegative it follows that economic evolution is directional and improves average economic performance provided that f and δ are the same for all firms.

Using the same tools of analysis we can also determine the properties of the selection equilibrium appropriate to a given technology set, a key determinant of which is the growth rate of the market environment. The solid lines m′ and g′ in Figure 4b show the equilibrium consistent with the given growth rate g_d. It is a logical requirement that any equilibrium must fall on the boundary of the selection set, and so the price of the index characteristic, v_o must adjust to support the equilibrium rate of growth. We can imagine an implicit demand curve for the chosen index characteristic which shifts at a rate g_d, and a supply of this characteristic from the various products expanding at a rate greater than this, to produce the required reduction in v_o. At the ruling market growth rate the dominant product in Figure 4b is β, the market share of which approach 100% asymptotically. However, at the ruling growth rate there is also a survival region, between m′ and g′, in which are located all other technologies which remain profitable but yet continue to decline in relative significance. It is thus vital to distinguish between survival and economic weight in a selection equilibrium. All products below m′ are eliminated during the course of the selection process. It follows immediately that a static environment, a zero value of g_d, eliminates this survival region, for the lines g′ and m′ now coincide in the line m″. Indeed with zero market growth, only one technology ultimately survives and this is γ which belongs to regime C, and not β. Thus the focal point of a selection process depends upon the market environment in two ways, via its growth rate and via its relative price structures which evaluate the competing products and processes. We can obviously interpret this selection process as bringing to market dominance a particular product design but it does not follow that the dominant design is, in an economic sense, the best-practice design. Indeed, if we interpret best-practice as applying to the product with the highest ratio of quality to unit cost of production, α_i/h_i, then this is represented by product γ which only comes to dominance in a static market environment. In a dynamic environment best-practice does not dominate the selection equilibrium, and in the context of some selection sets it may not even survive.

B. Some Extensions

Although we have focused on one particular highly simplified case to illustrate the method of analysis, this proves of great help in appraising various changes in assumption. The assumption of a non-uniform propensity to accumulate will serve to illustrate. To be specific, suppose that no firms are willing to grow except the firm with product k, which turns out to have the worst technology in the set. Despite this technological disadvantage, this firm is the only one which will grow and itself must come to account for 100% of the market in the selection equilibrium. All other products remain viable but their economic significance tends to zero as $\bar{h}$ converges on h_k and $\bar{\alpha}$ converges on α_k. In other words, once uniform propensities to accumulate are abandoned, there is no guarantee that inferior technology will not come to dominate the market. Evolution depends on other factors besides efficiency, and survival of the fittest does not entail that only the fittest survive.

We have so far found it convenient to present the argument in an excessively static form by holding the technology set fixed. In practice this set will change either through shifts in the price structure or though the development of new technologies. Such changes continually re-evaluate the selective force operating on a given product. Some of these changes may be due to exogenous events but others, particularly those relating to technology, may vary endogenously with scale economies, productive experience and creativity. There is no guarantee that a product which is best practice at one point in the selection process will remain in this favourable position throughout. Here, a great deal depends upon the agenda for change contained in the competing design configurations.

So it is important that this framework also permits one to judge the significance of new product and process innovations. We can immediately see that an innovation which does not redefine the boundaries of the technology set cannot come to dominate the market, and that the immediate success of an innovation depends on its location in the technology set. If located below 'm' (Figure 4a) it will be a failure from the outset. If located below 'm' and 'g' its output will grow but its relative market weight will decline. Only if located above 'g' will it enjoy some relative expansion from its entry position. Moreover, only if it lies North-West of β, Figure 4b, will it define a new best practice, dominant design. This does not imply, of course, that entry with a nonbest practice technology should be avoided. Quite the contrary, it may be highly profitable for a while, even though it is doomed eventually to relative decline or even absolute extinction.

Just as the entry of new technologies is an important element reshaping the selection process, so is exit. Although we cannot explore the issues here, we can show that maintaining production from inferior loss-making echnologies, in violation of the bankruptcy rule, must slow down the election process. Nor is this an unlikely event. Government subsidy, ccumulated capital reserves, or support from an umbrella organization an and do keep technologies 'alive' beyond their appropriate exit dates.

C. Creativity and Technology Strategy

We have not the space to explore in-depth the significance for the volutionary process of this third dimension of organizational perfornance. A few brief remarks must suffice. Innovation and creative change s a most difficult concept to handle and any formal expression of it may inder more than help. Nonetheless, as long as creative activity is reource dependent, it is clear that the operation of the selection process nust determine which firms are in a better position to enhance their echnological position. Consequently, those firms which do keep ahead f industry average performance will find their market share and resource ase continually expanding. So there must be some tendency for success o follow success, but there remain so many links in the chain between nowledge and wealth creation that this is only a statement of what is xpected to be true ceteris paribus.

Consider a firm operating within a design configuration that is sudenly enriched by outside developments in the knowledge base. It will be aced with a number of possibilities. It may not recognize the significance f the new developments, and soon find itself falling behind an industry verage state of technology which is improving rapidly to capitalize on ne new agenda for change. It may recognize the opportunity but be esource constrained and unable to advance its knowledge base at the equired rate. Without public subsidy, its only options may be to merge vith firms which possess the absent competences, to engage in a joint enture in collaborative technical development, or to license in the approriate technology. Finally, even with recognition and adequate resources, ne firm may still not keep its technological performance up with the ndustry trajectory. Poorly managed firms may yet prosper. Serendipity nay see to that. But once a firm falls behind the industry average its narket share and resource base will decline along the lines already indiated, so weakening still further its ability to stay within this line of roduction. Alternatively, it may capitalize successfully on the new agena for change and find itself in a position of market leadership and rowing cumulative advantage. These outcomes are, of course, impossile to predict but nonetheless the limits on the possible states of the world emerge clearly within the evolutionary framework. A great deal depend on a firm's expectations of the agenda for change, and thus on the desig configuration in which it operates.

Can anything more be said within the evolutionary context? One propo sition which we can establish here, concerns that direction of technologica advance that produces the maximum selective advantage for the firm; tha direction which is "optimal" in terms of the growth and profitability of th firm. Consider Figure 5 which relates to a particular firm, and which draw in part on Figure 4. This firm is currently located at point b, it is profitabl but it lies below average practice for the industry, represented by the line g g, which again has slope v_o^{-1}. On the basis of the resources available t this firm, it perceives that its technology may be improved anywher within its innovation opportunity set, those points bounded by the irregula curve z-z. The shape of this set suggests a rich design configuration wit three possible trajectories of change. Where should this firm aim to mov its technology? Clearly it should move to the boundary of z-z. Mor precisely, it should move to point b′, where the line g′-g′ is tangent to z-z for this is precisely the innovation which produces the greatest favourabl increase in the selective force operating on the firm. Now, it follow directly that the choice of this innovation trajectory is independent of th characteristics of the selective environment (the coefficients f and δ) an independent of the growth rate of the market environment. It does, how ever, depend on the price of the index characteristic v_o, since this determin es the relative selective advantage of product and process improvements Indeed, a lower value of v_o will be associated with a direction of chang which is biased toward cost reduction rather than quality improvemen This is shown by the dotted lines and the related point b″. Since we hav argued that v_o is reduced during the selection process, this result may we be consistent with the predictions of the product cycle literature (Utter back, 1979). Of course, unless firms have identical innovation opportunit sets they will not all end up with the same trajectories of innovation. No can identical opportunity sets be expected for the reasons outlined i Section III. Location in different design configurations, different percep tions of the agenda for innovation, different resource bases, and differer abilities to innovate will all create variety of innovative response, even fo firms within the same selection environment. Thus at the root of our entir argument lies variety in creativity, the fundamental fact which shapes th selection set, that fundamental fact which generates the possibility o selection.

From a technology strategy viewpoint this evolutionary framework

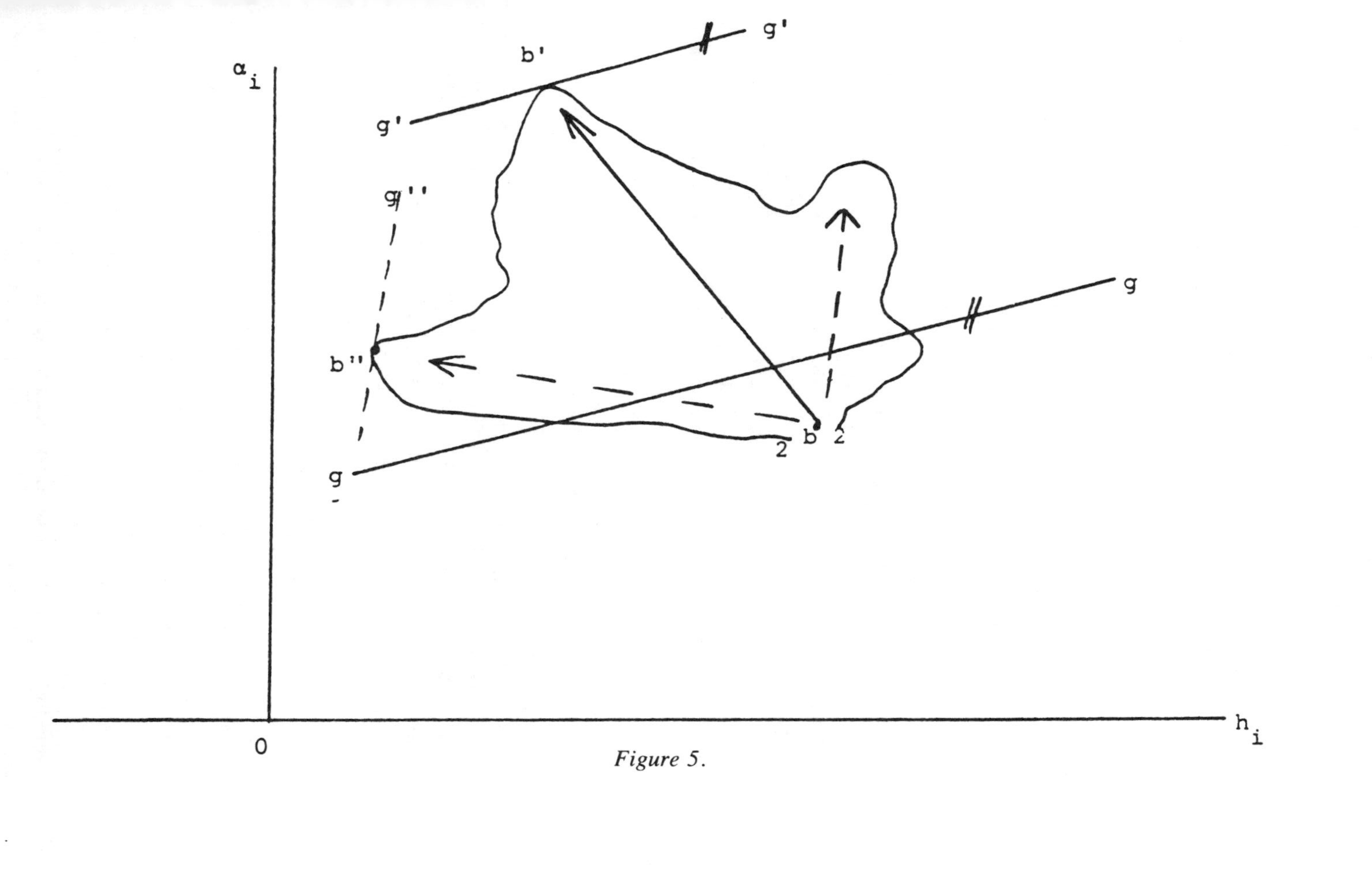

Figure 5.

raises interesting questions for any firm. Do management know the full extent of their technology set, how it is currently partitioned, and where they stand within it? Have they a correct perception of the characteristics which consumers value in the product and of their relative valuations? Are they aware of the imminent changes in this structure of economic valuations? Are they aware of the agendas for change contained within the different design configurations and of the inherent limits on revealed technological performance? Are they aware of new knowledge bases emerging to create rival technological regimes beyond the existing competencies of the firm? These are fundamental questions to be posed by a strategic management team, the answers to which may well differ between the strategists at business unit level and those at corporate level. Consider just three examples where these questions are of considerable current significance: as gallium arsenide develops as a rival for silicon semiconductors; as protein engineering threatens established chemical routes to the synthesis of pharmaceuticals and foodstuffs; and as ceramics provide the basis for entirely different approaches to engineering design. Within our evolutionary approach these questions fit naturally into place, since they emphasize quite explicitly the consequences of firms being different in their revealed technological performance and in their knowledge base. As history indicates, the sustained growth of a firm can depend a great deal upon its creativity and its ability to move into new configurations and even technological regimes. Those which do not make this difficult transition seem soon to fall into relative insignificance and even bankruptcy, as the recent history of the microelectronics industry so well illustrates (Braun and McDonald, 1978; Malerba, 1985; Foster, 1986). Thus knowledge and the epistemology of the firm lie at the very basis of the modern competitive process.

CONCLUSIONS

The central theme of this essay has been that economic change is driven by economic variety, and that the chief long run source of economic variety is to be found in technological innovation. We have argued that to analyze these phenomena requires a major change in thinking, away from the search for ideal, representative types which can only mask the forces which drive progress and change, towards the search for distributions of phenomena. In our proposed structure, organizations play the key role in generating and applying useful knowledge to the economic process: they are the proximate sources of variety, carrying technological performance through time in a structured, purposive way. The manner in which they achieve this role is both constrained and stimulated by their environment. We believe that this is a rich framework of analysis, capable of encompassing the wide variety of ways in which organizations articulate technology, and yet capable of giving that variety causal significance in the process of change. We hope this essay will stimulate others to develop critically the evolutionary perspective around the triad of technology, organization and environment.

ACKNOWLEDGEMENT

We are particularly grateful to our colleagues in PREST, L Georghiou, Janet Evans, Tim Ray, Hugh Cameron, Brendan Barker, Mark Boden and J. J. Chanaron who have helped shape our views on technology and competition. The contribution of the editors to this final version is gratefully acknowledged, as are the insightful comments of Professors Brian Loasby and Ian Steedman. Needless to add they bear no responsibility for the final outcome. Financial support from the ESRC programme on Competitiveness of UK industry is also gratefully acknowledged.

REFERENCES

Abernathy, W., 1978. *The Productivity Dilemma,* Johns Hopkins, Baltimore.

Aldrich, H. and Mueller, S., 1982. "The Evolution of Organizational Forms." *Research in Organizational Behaviour,* Vol. 4.

Borrus, M., Millstein, J. E., and Zysman, J., 1983. "Trade and Development in the Semi Conductor Industry: Japanese Challenge and American Response," in Zysman, S. and Tyson, L. (eds.), *American Industry in International Competition,* Cornell University Press.

Braun, E. and McDonald, S., 1978. *Revolution in Miniature.* Cambridge University Press.

Burns, T. and Stalker, G. M., 1961. *The Management of Innovation.* Tavistock, London.

Cooper, A. C. and Schendel, D., 1976. "Strategic Responses to Technological Threats," *Business Horizons,* Vol. 19, pp. 61–69.

Dosi, G., 1983. "Technological Paradigms and Technological Trajectories," in C. Freeman (ed.), *Long Waves in the World Economy,* Butterworth.

Downie, J., 1955. *The Competitive Process.* Duckworth, London.

Foster, R., 1986. *Innovation.* Macmillan, London.

Frankel, M., 1955. "Obsolescence and Technological Change," *American Economic Review,* Vol. 45, pp. 296–319.

Georghiou, L., Metcalfe, J. S., Evans, J., Ray, T., and Gibbons, M., 1986. *Post Innovation Performance.* Macmillan, London.

Gilfillan, S., 1935. *Inventing the Ship.* Follett, Chicago.

Gould, S. J., 1980. "The Return of Hopeful Monsters." Reprinted in *The Panda's Thumb,* Penguin, London.

Hannan, M. T. and Freeman, J., 1977. "The Population Ecology of Organizations." *American Journal of Sociology,* Vol. 82, pp. 929–964.

Hayek, F., 1948. *Individualism and Economic Order,* Chicago University Press.

Hrebiniak, L. G. and Joyce, W. E., 1985. "Organizational Adaptation: Strategic Choice and Environmental Determinism." *Administrative Science Quarterly,* Vol. 30, pp. 336–349.

Itami, H., 1987. *Mobilizing Invisible Assets.* Harvard.

Jantsch, E., 1967. *Technological Forecasting in Perspective,* OECD, Paris.

Jurkovitch, R., 1974. "A Core Typology of Organizational Environments." *Administrative Science Quarterly,* Vol. 19, pp. 380–394.

Lawrence, P. R. and Lorsch, J. W., 1967. *Organization and Environment,* Harvard University.

Layton, E. T., 1974. "Technology as Knowledge." *Technology and Culture,* pp. 30–41.

Levins, R., 1968. *Evolution in Changing Environments,* Princeton.

Loasby, B., 1976. *Choice, Complexity and Ignorance.* Cambridge: Cambridge University Press.

Loasby, B., 1982. "The Entrepreneur in Economic Theory." *Scottish Journal of Political Economy,* Vol. 29, pp. 235–245.

Matthews, R. C. O., 1985. "Darwinism and Economic Change," in Collard, D. et al. (eds.), *Economic Theory and Hicksian Themes,* Oxford University Press.

Marshall, A., 1920. *Principles of Economics 8th edition.* Variorum, MacMillan.

Malerba, F., 1985. *The Semiconductor Business.* F. Pinter, London.

Mayr, E., 1982. *The Growth of Biological Thought.* Harvard University Press.

McElvey, W., 1982. *Organizational Systematics.* University of California Press.

McNulty, P. J., 1968. "Economic Theory and the Meaning of Competition." *Quarterly Journal of Economics,* Vol. 82, pp. 649–656.

Metcalfe, J. S. and Gibbons, M., 1986. "Technological Variety and the Process of Competition," *Economie Applique,* pp. 493–520.

Millstein, J. E., 1983. Decline in an Expanding Industry: Japanese Competition in Color Television, in Zysman, S. and Tyson, L. (eds.), *American Industry in International Competition,* Cornell University Press.

Monod, J., 1971. *Chance and Necessity,* Penguin, London.

Morgenstern, O., 1972. "Thirteen Critical Points in Contemporary Economic Theory." *Journal of Economic Literature,* Vol. 10, pp. 1163–1189.

Nelson, R. and Winter, S., 1983. *An Evolutionary Theory of Economic Change.* Harvard University Press.

Penrose, E., 1959. *The Theory of the Growth of the Firm.* Blackwell, London.

Phelps, E. and Winters, S., 1970. "Optimal Price Policy Under Atomistic Competition," in E. Phelps (ed.), *Mirco Foundations of Employment and Inflation Theory,* W. W. Norton, New York.

Richardson, G. B., 1972. "The Organization of Industry." *Economic Journal,* Vol. 82, pp. 883–896.

Rosenbloom, R. S., and Abernathy, W., 1982. "The Climate for Innovation in Industry," *Research Policy,* Vol. 11, pp. 209–225.

Rosenberg, N., 1982. *Inside the Black Box: Technology and Economics.* Cambridge University Press.

Sahal, D., 1981. *Patterns of Technological Innovation.* Addison, Wesley, New York.

Saviotti, P., 1984. "Indicators of Output of Technology," in M. Gibbons et al., (eds.), *Science and Technology Policy in the 1980s and Beyond,* Longman, London.

Silverberg, G., 1985. "Technical Progress, Capital Accumulation, and Effective Demand," in D. Batten (ed.), *Economic Evolution and Structural Change,* Springer-Verlag, Berlin.

Sober, E., 1985. *The Nature of Selection,* MIT Press, London.

Teece, D., 1987. "*Technological Change and the Nature of the Firm,*" mimeo, Los Angeles, CA: University of California, School of Business Administration.

Terreberry, S., 1968. "The Evolution of Organizational Environments," *Administrative Science Quarterly,* Vol. 13, pp. 590–613.

Tushman, M. L. and Anderson, P., 1985. "Technological Discontinuities and Organizational Environments," *mimeo,* New York: University of Columbia Business School.

Turner, R., and Soete, L., 1984. "Technology, Diffusion and the Rate of Technological Change." *Economic Journal,* Vol. 94, pp. 612–623.

Utterback, J. M., 1979. "The Dynamics of Product and Process Innovation in Industry," in Hill, C. T., and Utterback, J. M. (eds.), *Technological Innovation for a Dynamic Economy,* Elmsford, N.Y.: Pergamon.

Parijs, P. Van, 1981. *Evolutionary Explanation in the Social Sciences.* Tavistock, London.

Phelps, E., and Winter, S., 1970. "Optimal Price Policy Under Atomistic Competition" in Phelps, E. (ed.), *Micro Economic Foundations of Employment and Inflation Theory,* Norton, New York.

De Vincenti, W., 1984. "Technological Knowledge Without Science: The Innovation of Flush Riveting in American Airplanes," *Technology and Culture,* Vol. 25, pp. 540–576.

Williamson, O. E., 1986. *Economic Organization: Firms, Markets and Policy Control.* Wheatsheaf, London.

Winter, S., 1982. "An Essay on the Theory of Production," in S. H. Hyams (ed.), *Economics and the World Around It.* Ann Arbor, MI: University of Michigan Press.

A FRAMEWORK FOR UNDERSTANDING THE EMERGENCE OF NEW INDUSTRIES

Andrew H. Van de Ven and Raghu Garud

ABSTRACT

Knowledge about the process by which new industries emerge is invaluable both to industrial policy makers and to corporate managers and entrepreneurs. This chapter proposes a framework for viewing an industry as a social system, and adopts an accumulation theory of change to examine industry emergence. The framework examines the processes by which industries emerge over time, as well as the roles of individual firms in creating an industry.

Research on Technological Innovation, Management and Policy
Volume 4, pages 195–225.

ISBN: 0-89232-798-7

INTRODUCTION

How do new industries emerge? What are the roles of individual firms in creating an industry? These questions not only have significant implications for national industrial policy, they are critical to innovation managers. The purpose of this chapter is to introduce a framework and a theory of change for understanding the emergence of industries and the roles of individual firms within it to commercialize "technological" innovations. Seldom can such technological innovations be developed by a single firm alone in the vacuum of a community or industrial environment. Many complementary innovations are usually required before a particular technology is suitable for commercial application (Rosenberg, 1983, p. 56). Reviews by Mowery (1985) and Thirtle and Ruttan (1986) clearly show that the commercial success of a technological innovation is in great measure a reflection of institutional innovations which embody the social, economic, and political infrastructure that any community needs to sustain its members. Thus, the management of innovation must not only be concerned with micro developments of a particular technical device or product, but also with the creation of an industry infrastructure needed to commercialize the innovation.

While most innovations represent small increments of normal change that refine and improve an established order, this chapter examines "extraordinary innovations" (Rosenbloom, 1985) which set in motion a sequence of events that can disrupt, destroy, and make obsolete established competence, or create totally new organizations and industries. In other words, "these innovations have a 'transilient' . . . capacity to transform established systems of technology and markets" (Abernathy and Clark, 1983, p. 13).

For example, in November 1984, the U.S. Food and Drug Administration (FDA) announced its approval for commercial release a cochlear implant, which is a biomedical device that permits many profoundly deaf people the ability to discriminate sound. In its announcement, the FDA reported that it was the first time that one of the five human senses has been replaced by an electronic device. In October 1985, the FDA approved a second implant device (Yin and Segerson, 1986). The cochlear implant represents an extraordinary innovation that has the potential of transforming the traditional hearing-aids industry, which serves only individuals with impaired hearing, not the profoundly deaf. The commercialization of cochlear implants requires the development of a totally new set of skills, knowledge, and institutional arrangements. To date, these include the development of new diagnostic and surgical procedures, ontological service facilities, trained technicians, as well as the functional competencies of R&D, manufacturing, and marketing. Commercialization of cochlear implants also requires the creation of new industry practices and FDA regulations and standards of efficacy and safety of the devices.

Part I of this chapter examines the traditional definition of an industry and the theory of natural selection that has been used to explain industrial change by industrial economists and organization theorists. We will explain why this definition is too narrow, and why this theory of natural selection cannot explain the creation of new industries. Part II will propose an accumulation theory of change and a framework for viewing an industry as a social system. This framework is broader than most treatments of industries, and combines relevant perspectives from industrial economics, marketing channels, and organizational sociology. In Part III we will examine key processes of industry emergence, both from a macro perspective of the industry as a whole, as well as from a micro perspective of individual firms constituting the industry. Finally, a concluding section will discuss the strengths and limitations of the social system framework for studying industry emergence over time.

PART I: TRADITIONAL FRAMEWORKS OF INDUSTRY EMERGENCE

Two issues are basic to understanding industry emergence: (1) definition of an industry, and (2) a theory of how industries change. A review of how these issues have been addressed by industrial economists and organization theorists provides the conceptual foundation for proposing that an industry be viewed as a social system.

A. Industry as a Group of Naturally Selected Competitors

Industrial economists typically define an industry as "the group of firms producing products that are close substitutes for each other" (Porter, 1980, p. 5). They adopt a natural (economic) selection theory of change to explain how industry structures (ranging from perfect competition to monopoly) are transformed from within by the differential success of competing members (Kamien and Schwartz, 1982, p. 2). As Porter

(1980) discusses, economic competition works to drive down the rate of return on invested capital toward a perfectly competitive floor rate of return. Investors whose long-term returns are below this "free market" return will invest elsewhere, and firms habitually earning less than this return will eventually go out of business. Rates of return higher than this floor rate will stimulate the inflow of capital into an industry either through new entrants or through additional investment by existing competitors.

This industrial economics view of industry change is fundamentally the same as the natural selection theory used in the population ecology model of organizations. The analysis typically begins with a given population of organizations that are mutually susceptible to environmental vulnerability (Hannan and Freeman, 1977). These firms share a "commensalistic interdependence," which implies that they engage in a competitive "intraspecies" struggle for economic survival because they are commonly subject to the same environmental fate (Van de Ven and Astley, 1981, p. 444). Porter describes this common environmental fate as consisting of five economic forces: firm entry and exit barriers, threats of substitution, bargaining power of buyers and suppliers, and rivalry among current competitors (Porter, 1980).

Because an industry is viewed as consisting of an agglomeration of firms producing similar or substitute products, natural or economic selection drives member firms over time to become increasingly homogeneous within clusters of "strategic groups" (Porter, 1980) by adopting some common key elements, such as technology, knowledge, organizational forms and practices. Membership to groups is restricted by the erection of entry and mobility barriers. Within each group, natural selection incrementally and gradually transforms an industry from within through a one-by-one selection of industry participants over time. Some organizations fail and are selected out, while others survive competitive economic forces (Astley and Van de Ven, 1983). At the same time, new organizations are created and enter the industry. As these firms replace their failed predecessors, the industry as a whole gradually changes composition.

Sahal (1981), Dosi (1982), and Rappa (1987) note that the core technologies on which industries are founded persist. The basic design of a technology embodies a set of ground assumptions and starting premises that constrain the course of subsequent improvements in an industry and govern the extent to which further innovation is possible. In particular, Sahal cites farm tractors, airplanes, and electric motors as examples of industries that relied on technologies introduced over 50 years ago. While these technologies have undergone significant cost-reduction improvements since then, they have occurred only through gradual refinements of the same basic technologies.

In support of his "principle of technological insularity," Sahal (1981, p. 57) observed a lack of inter-industry transmission of technical know how. In large part, know-how seems to be specific to an industry (Rappa, 1987). The development of technology takes place through a process of learning that is context dependent and largely self-contained in the industry of its origin. "Since learning takes place through direct experience, industries face significant costs not only in the search for technologies developed elsewhere, but in adapting those technologies within the industry" (Astley, 1985, p. 227). Such costs drastically reduce the probability that significantly new techniques will penetrate an industry and limit the pool of variations within an industry to small deviations from established norms. Consequently, selection from this restricted range of variations inevitably moves an industry to stabilize its form over time.

As Astley (1985) argues, by concentrating on natural "selection" (as distinct from "variation") processes within an increasingly homogeneous industry, the economic population ecology perspective cannot explain either the formation of entirely new industries or explain major extensions of old industries that are brought on by technological innovation. Paradoxically, environmental selection works best in industries saturated with competitors because under conditions of resource scarcity only the firms that compete best for limited resources survive. But these same conditions encourage homogeneity and uniformity within industries and effectively inhibit rather than promote industry change because selection only works to bring about change when sufficient diversity or variations exist.

B. Creative Destruction and Punctuated Equilibrium Theory of Change

Schumpeter (1975) pointed out this basic incompatibility between entrepreneurial activity and perfect competition, and argued that while the latter can assure efficient allocation of resources at every point in time, it stifles the inventive activity that most efficiently allocates resources over time. Envisioning an ever changing (as opposed to a stationary equilibrium) economic environment, Schumpeter argued that the concept of perfect competition is irrelevant because it focused entirely on *market (price) competition,* when the focus should be on *technological competition.* What counts, Schumpeter (1975, p. 84) stated is:

> competition from the new commodity, the new technology, the new source of supply, the new type of organization (the largest-scale unit of control for instance)—competition which commands a decisive cost or quality advantage and which strikes not at the margins of the profits and outputs of the existing firms but at their foundations and their very lives.

Instead of relying on market prices as the selection mechanism of technological directions in the initial stage of industry emergence, Schumpeter stressed technological competition among a multiplicity of risk-taking actors who are involved in a trial-and-error search process of commercializing new technologies. Competition does not only occur between an entrepreneur's "new" technology and the "old" substitute technology, but also among alternative "new" technological approaches being pursued by other entrepreneurs (Dosi, 1982, p. 155). These actors take risks, of course, because there are markets which allow high rewards (i.e., profits) in cases of commercial success.

Schumpeter justified certain monopolistic practices because in the course of introducing a new commercial innovation a firm must be permitted to engage in monopolistic practices designed to retard imitation long enough for the firm to reap monopoly profits from its investment. Since monopoly positions are temporary, the process of creative destruction benefits society through the introduction of new products or processes, and by recognizing that the extraordinary profits realized by the innovating firm will be a source of funds for the next round of innovation. As examples of this process, Kamian and Schwartz (1982, p. 10) cite ball-pens replacing fountain pens, jet-engined aircraft replacing piston engines, pocket calculators replacing slide rules, and electronic watches replacing mechanical ones.

Unlike price competition, where rivals compete by producing standardized goods at a lower price in saturated markets, technological competition operates by discovering new, unsaturated niches and offering innovative goods that no one else has supplied. The initial existence of untapped demand encourages experimentation and permits clusters of entrepreneurs pursuing new technological variations to emerge and coexist before a dominant form is competitively selected in later stages of industry growth (Astley, 1985, p. 230).

A punctuated equilibrium model of change (as borrowed from Gould and Eldredge, 1977) appears more appropriate for describing this process of creative destruction and technological competition than the continuous and incremental evolutionary model of natural selection discussed above. In a "punctuated equilibrium" model divergent technological innovations introduce relatively short periods of discontinuous ruptures in between extended periods of convergent incremental change of a unit. In other words, the random selection of a major technological variation "punctuates" extended periods of continuous equilibrium producing improvements.

For example, Piore and Sabel (1984) discuss how long periods of stability are abruptly ended by innovative breakthroughs that move industrial evolution down entirely new paths. They indicate that stability and continuity result from the adoption of a given technology because it typically entails large investments in equipment and know-how, which in turn discourage subsequent changes in industrial development. But eventually, developments with this technology become marginal and/or a mutant technology appears that fosters the emergence of a new industrial sector. These extraordinary innovations are often rejected by an existing industry because of the strength and inertia built into its existing technological paradigm. Analogous to a scientific paradigm, Dosi (1982, p. 153) and Rappa (1987) argue that technological paradigms have a powerful exclusion effect; the efforts and the technological imagination of engineers are focused in rather precise directions while they are, so to speak, "blind" with respect to other technological possibilities. As a consequence, entrepreneurs often have no recourse but to isolate themselves from existing industry branches or entire industries and "start from scratch" to create a new industry infrastructure that is necessary to commercialize their technological innovation (Dosi, 1982, p. 154).

But an explanation of this punctuation process requires a more encompassing definition of an industry and a more fine-grained theory of industry change than the punctuated equilibrium model provides.

First, similar to the concept of an organization field proposed by institutional theorists (DiMagggio and Powell, 1983; Granovetter, 1985), the traditional definition of an industry needs expansion not only to include competing firms, but also many other actors that perform all the functions necessary to develop and commercialize a technological innovation. Although these other organizations and functions have heretofore been treated as "externalities" (Porter, 1980), they are crucial actors and functions to incorporate directly in an explanation of industry emergence. In the next section we will propose a social system framework that provides this needed broader perspective of an industry.

Second, while the punctuated equilibrium model is helpful for describing the overall life cycle of an industry, it does not adequately explain the process of "punctuation" itself. So also, while Astley (1985) rightly criticizes the overemphasis by population ecologists on "selection" processes when an understanding of change calls attention to the process of "variation," he does not explain how variations emerge and grow. Yet, it is precisely this process of "punctuation" or "variation" that needs to be understood to explain the process of industry emergence. As Rosenberg (1983, p. 60) suggests, even spectacular technological breakthroughs require many complementary technological and institutional innovations before they are ready for commercial application. Therefore, all the activities and events involved in the development and commercialization of such extraordinary technologies do not take place "instantaneously." It is a process that unfolds over a considerably long period of time, and often through the accumulation of many small improvements that make the eventual commercial application of the extraordinary technology possible. An *accumulation theory of change* appears more appropriate for explaining industry emergence than either the punctuated equilibrium model or the evolutionary theory of natural selection.

PART II. AN INDUSTRY AS AN EMERGING SOCIAL SYSTEM

As we have discussed, a technological innovation represents a new technological trajectory (Nelson and Winter, 1982) that diverges from and is often rejected by an existing industry because of its discontinuous rupturing effects on the established technological paradigm. Thus, to explain how this new technological trajectory develops one has to start almost from the beginning to explain how a new industry is born. An *accumulation* or *epigenetic* theory of change (as described by Etzioni, 1963) appears particularly appropriate for explaining industry emergence. We will apply this accumulation theory of change at two levels of analysis: (1) the motivations and activities of individual firms or entrepreneurs, and (2) the collective level of multiple actors who interact and socially construct an industry.

At the level of individual entrepreneurs, the accumulation theory begins with the purposeful intentions and business ideas of entrepreneurs, and posits that a totally new unit (be it a new technology, business venture, or program) is created through a stream of activities undertaken by entrepreneurs to accumulate the external resources, competence, and ingredients necessary to transform and construct the business ideas into a self-sustaining economic enterprise. Concepts of initiation, takeoff, and startup are important for describing this accumulating change process (Van de Ven and Poole, 1987). *Initiation* is the time when entrepreneurs decide to form a business venture (if successfully launched will become the birthday of the business unit), and *takeoff* is the time when the unit can exist without the external support of its initiators and continue growing "on its own." The period between initiation and takeoff could be called *startup,* where the new unit must draw its resources, competence, and technology from the founding leaders and external sources in order to develop the proprietary products, create a market niche, and meet the institutional standards established to legitimate the new unit as an ongoing economic enterprise.

Thus, the accumulation model views change as being stimulated by external forces during the initiation period, a transition from external to internal sources of change during the startup period, culminating at takeoff with the capability for immanent development. In comparison with the other change theories previously discussed, the initiation and startup periods correspond to the "punctuation" period of technological competition, and the period after takeoff in the accumulation model converges into an incremental process of increasing refinements and evolutionary change as described by the natural selection model of price competition.

The process of accumulation is like that of an airplane that first starts its engines and begins rolling, still supported by the runway, until it accumulates enough momentum to "take off," and continue in motion "on its own" energy to carry it to higher altitudes and speeds. Thus, while relying initially on external support, the necessary condition for autonomous action is produced through a process of accumulation.

Etzioni (1963, p. 490) emphasizes that important relations of *timing, isolation, momentum, and critical mass* exist between initiation and takeoff points. For example, the longer, more isolated, and incremental the startup period, the lower the market awareness and interest in the new venture, and the more difficult it is to mobilize commitment of investors and involvement of potential key suppliers and customers. On the other hand, if the startup period is too compressed, there may be insufficient time during startup to establish competencies, develop quality products, or build the momentum and critical mass needed for market penetration to sustain the new venture at takeoff.

At the aggregate system level, industry emergence represents the cumulative achievements of a new ''community'' of symbiotically-related firms and actors who, through individual and collective action, invest resources in and transform a technological invention into a commercially viable business. Beginning with the initiation period, paths of independent entrepreneurs (acting out their own diverse intentions and ideas) intersect. These intersections provide occasions for interaction. Through interactions the actors come to recognize areas of symbiotic and commensalistic interdependence. Symbiosis (as opposed to commensalism) promotes cooperation because actors can achieve complementary benefits by integrating their functional specializations. Out of these interactions there emerges a social system of organizational units that increasingly isolate themselves from traditional industries by virtue of their interdependencies, growing commitments to, and unique know-how of a new technology trajectory. This quasi isolation signifies the beginning of the start-up period of the emerging industry. Isolation frees the new system from institutional constraints of earlier industries, and permits it to develop its own distinctive structural form (Rappa, 1987).

The emergence of the cochlear implants industry provides an illustration of how the cochlear implant community achieved isolation from the existing hearing aids and vibrotactile community (Garud and Van de Ven, 1989). Two mechanisms by which this isolation was achieved will be provided here as examples. First, isolation was achieved by the accumulation of a critical mass of people who believed that the cochlear implant technology could provide benefits that existing technologies could not provide. As this number of believers increased, the number of activities (such as technical conferences and training programs) that shared and promoted developments in cochlear implant technology increased manifold, culminating in the endorsement of cochlear implants as a useful procedure by the American Medical Association in 1983.

Second, isolation was achieved by creating new institutional forms and mechanisms. For instance, a separate panel of evaluators for cochlear implants was established within the U. S. Food and Drug Administration that allowed cochlear devices to be evaluated on new terms and standards, rather than on the basis of standards employed for evaluating distant substitute products such as vibrotactile devices and hearing aids. Similarly, in 1984, the American Speech and Hearing Association established a special committee on cochlear implants which began evaluating and reporting on aspects concerning the safety and efficacy of cochlear devices with new terms and standards.

During this initial startup period, ''structuration'' of an organizational field occurs as a result of the activities of a diverse set of organizations (DiMaggio and Powell, 1983). The process of structuration consists of four parts: (1) an increase in the extent of interaction among organizations in the field, (2) the emergence of sharply defined interorganizational structures of domination and patterns of coalition, (3) an increase in the information load with which organizations in a field must contend, and (4) the development of a mutual awareness among participants in a set of organizations that they are involved in (DiMaggio, 1982). Thus, coordination among actors takes place not so much by a central plan or organizational hierarchy, nor through the price mechanism, but mostly through interactions (Mattsson, 1987) and partisan mutual adjustments among actors (Astley and Van de Ven, 1983). After all, environmental niches do not pre-exist, waiting to be filled (McKelvey, 1982, p. 109); they are socially constructed domains through the opportunistic and collective efforts of interdependent actors in common pursuit of a technological innovation.

While price competition is largely absent in the industry startup stage (since no products have been introduced in the market), technological competition emerges as alternative technological paths become evident and different entrepreneurs or firms ''place their bets on'' and pursue one or more of these alternatives. It must be emphasized that applied R&D is highly uncertain and dependent in this startup period. A number or risk-taking actors emerge, ready to develop different technical and commercial solutions. Moreover, depending on the technological alternative chosen by a firm, it becomes highly dependent upon different clusters of institutions (universities, laboratories, disciplines) which have been producing and directing the accumulation of basic knowledge, techniques, and experience associated with a given technological alternative. For instance, during the early stage of cochlear implant industry development, five different technological routes were taken by five different business firms for development and commercialization. Each firm became dependent on a different cluster of research institutes and universities depending on the particular technological route that they had chosen to pursue.

As the number of organizational units and actors gains a critical mass, a complex network of cooperative and competitive relationships begins to accumulate. This network itself becomes recognized as a new ''industrial sector,'' and takes on the form of a hierarchical,[1] loosely-joined system. We view this emerging system as consisting of the key firms and actors that govern, integrate, and perform all of the functions required to transform a technological innovation into a commercially viable line of products or services delivered to customers. We will describe the structure of this system. When fully developed, it consists of a number of subsystems, and each subsystem performs a limited range of specialized functions.

A. The Structure of an Industry Social System

Granovetter (1985) suggests that social systems are neither under socialized nor over socialized. They are in fact, embedded in social relationships that emerge from the interactions of the organizations constituting the system. These social systems are hierarchically stratified with different subsystems specializing in different functions—technical instrumental functions, resource procurement functions, and institutional legitimation and governance (Parsons, 1964).[2] Although these functional subsystems are highly related, we will distinguish them here to provide analytical guidance for investigating the emergence of an industry. Figure 1 provides an overview of these subsystems.

1. Instrumental Subsystem

The instrumental subsystem incorporates the traditional industrial economic definition of an industry, and consists of the firms producing products that are close substitutes for each other. The focus here is on the individual entrepreneurs and firms that begin applied R&D work in areas related to a technological innovation. If they persist in developing the

Subsystem	*Major Functional Activities*
Institutional	Industry governance structure Industry legitimation and support Industry rules, regulations and standards
Resource Procurement	Basic scientific or technological research Financing, insurance, and venture capital Human resources competence and accreditation
Instrumental	Commercial product development by proprietary firms, vendors, suppliers of: applied R&D, manufacturing, marketing, distribution

Figure 1. Functional Activities in an Industry Social System

innovation, they subsequently develop a line of products and functional competencies (e.g., manufacturing and assembly, marketing and distribution, etc.) to establish an economically viable business.

Williamson's (1975, 1979) transactions costs theory is useful for understanding how firms organize to perform these economic instrumental activities. He argues that those instrumental activities that are recurrent, entail idiosyncratic investments, and are highly uncertain to execute are performed within competing firms, while those that tend to be nonspecific, occasional, and for which there are a number of supply sources are licensed or contracted by firms with outside suppliers and vendors. However, in the initial stages to commercialize a technological innovation, many different inputs belonging to different team members are used and their activities are often indivisible in that they do not yield identifiable, separate transactions which can be summed to measure the total output. It is these conditions requiring "team productive process" that explain the emergence of the firm as distinct from market arrangements (Alchian and Demsetz, 1972).

From a system perspective, these make or buy decisions by individual firms produce the aggregate industry channels of raw materials, manufacturing, marketing and distribution flows (Stern and El-Ansery, 1982). Shifts in individual firm make-or-buy decisions over time also determine changes in industry structure. Stigler (1957) examined this issue in terms of the relationship between product life cycle and vertical integration. He argued that early in the product life cycle, vertical integration of functions within firms (i.e., make decisions) will predominate because the market will not yet have had a chance to develop. As Smith (1776) indicated, the division of labor among firms is limited by the extent of the market. When the market expands, many of the tasks and functions grow into sufficient size to make it more profitable to "buy" them from specialized manufacturers or suppliers, as opposed to "making" them within vertically-integrated firms.

2. *Resource Procurement Subsystem*

This subsystem of an industry includes the basic resources necessary to support proprietary instrumental activities. Akin to the classical economic resources of land, labor, and capital, three basic kinds of resources are critical to the emergence of most every industry: (a) basic scientific or technological knowledge, (b) financing, and (c) a pool of competent human resources (Mowery and Rosenberg, 1979). While private firms do engage in the development of these resources, typically public organizations (often viewed as "external" to an industry) play a major role in creating and providing these "common goods."

a. *Basic scientific or technological research* provides the foundation of knowledge that makes the commercial birth of an industry possible. But this basic knowledge is very costly to produce relative to its cost of diffusion and imitation (Mansfield, 1985). In addition, it builds in a cumulative fashion, and its generation is inherently an indivisible activity (Metcalfe and Soete, 1983). For these reasons, Nelson (1959) and Arrow (1962) argued that the social returns to research investment exceed the private returns to individual firms; a condition leading to underinvestment by the firm (from the social point of view) in research. As a consequence, a variety of studies have shown that firms rely upon outside sources of knowledge and technical inventions for the vast majority of their commercially significant new products (Rosenbloom, 1966; Mueller, 1962; Utterback, 1974; Stobaugh, 1985).

> In most industries, no single firm commands a majority of the resources available for research, nor can any one firm respond to more than a portion of the needs or problems requiring original solutions. It is not surprising, therefore, to find that most of the ideas successfully developed and implemented by any firm came from outside that firm (Utterback, 1974, p. 621).

b. *Financing*. While public institutions (e.g., NSF) tend to play the major role in financing the development of basic scientific or technological knowledge, venture capital (either within a corporation or in the market) tends to be the key financial source that supports private firms to transform this basic knowledge into proprietary and commercial applications. In addition, the commercialization of many technological innovations require unique industry-wide financing arrangements. For example, few bio-medical innovations would be commercially viable without the health care insurance industry and the creation of third-party payment reimbursement systems. Without such a financial infrastructure for a broad array of bio-medical and health care innovations, most patients would be unable to pay for many bio-medical devices and treatments. But since these insurance systems limit coverage to specifically-designated medical devices and treatments, the firms competing to commercialize a specific bio-medical device must cooperate to both educate and influence third-party payors to include the innovation in their payment reimbursement systems.

c. *A pool of competent human resources* is another essential resource necessary for the emergence of a new industry. New technologies mean new ways of performing essential tasks, be they related to research, manufacturing or marketing. This pool of competence tends to develop in three ways. First, firms involved in instrumental activities recruit and train people in specific skills related to the innovation. Over time, such professionals could leave the parent firm to join others, thereby diffusing such skills throughout the industry. A second method of developing this labor market is through the establishment of industry conferences, technical committees, trade publications, and technical journals, which provide an opportunity for industry participants to share and learn from each other. Finally, the competence pool is created through "collective invention" (Allen, 1983). In an interesting examination of cooperative R&D among rivals, von Hippel (1986) observed extensive trading of proprietary know-how among informal networks of process engineers in rival (and nonrival) firms in the U.S. steel minimill industry and elsewhere. He reasons that the sharing of proprietary competence and know-how among competing firms is most effective when:

> (1) the needed know-how exists in the hands of some member of the trading network, and when (2) the know-how is proprietary only by virtue of its secrecy, and when (3) the value of a particular traded module is too small to justify an explicit negotiated agreement to sell, license or exchange (von Hippel, 1986, p. 26).

Since advances in knowledge among a network of technological innovators often consist of small, incremental pieces of information, von Hippel's analysis suggests that the universe bounded by these three conditions is likely to have a substantial impact in building a competence pool among rivals and noncompeting firms.

3. *Institutional Subsystem*

The ultimate authorities governing and legitimating collective action are the rules and norms of the society in which organizations function (Galaskiewicz, 1985). The political context is the place for formally institutionalizing and legitimating a social system, which permits it to operate and gain access to the resources it needs (Pfeffer and Salancik, 1978, p. 214). As discussed below, institutional functions include: (1) establishing governance structures and procedures for the overall industry, and (2) legitimizing and supporting the industry's domain in relation to other industrial, social, and political systems. The success or failure of a new industry and firms within it depends on their abilities to achieve institutional isomorphism (Dimaggio and Powell, 1983). To achieve this, firms may either adapt to institutional requirements or attempt to build their goals and procedure directly into society as institutional rules (Meyer and Rowan, 1977). Thus, firms compete not only in the marketplace, but also in this political institutional context. Rival firms often cooperate to collectively manipulate the institutional environment to legitimize and gain access to resources necessary for collective survival (Pfeffer and Salancik, 1978; Hirsch 1975; Miles and Cameron, 1982).

a. Governance. The potential for domination and coercion by powerful firms over powerless actors is circumscribed by norms, customs, and laws of reasonable value and practice, which are the "working rules" of collective action (Commons, 1950). Norm formation results from collective interaction. In the process of becoming dissociated with their specific origins and transmitted through customs and laws, norms become generally applicable sources of collective regulation (Durkheim, 1933). Though they arise from the expedient acts of firms, when society views them legitimate, they become regulative, take on social force, and become the "institutionalized thought structure" for directing and regulating industrial life (Warren, 1971).

It is widely recognized that a variety of governmental regulations and institutional arrangements facilitate and inhibit the emergence of new technologies and industries. Mowery (1985) for example, discusses how government funding, by broadening the industry-wide knowledge base, can encourage new industry entrance, and thereby support a more competitive environment. So also, a more permissive antitrust policy permitting certain kinds of joint research ventures among competitors, as well as current requirements by the Department of Defense for licensing and second sourcing of new devices, speed the diffusion of innovation and may aid the operation of competitive forces in an industry. However, institutional policies encouraging rapid knowledge diffusion, if pursued too eagerly, may undermine the return to the knowledge producer, and thus the incentive to invest in information producing activities. But here another institutional mechanism has been devised; the patent system grants monopoly rights to the use of knowledge for a limited period of time. Although these institutional arrangements are often highly imperfect, it is clear they often exert a profound effect on technological and industry development.

b. Legitimation. Trust, or lack of customer uncertainty about product quality, is fundamental to efficient operation of the market mechanism. Under conditions of high quality uncertainty, "lemons" (or inferior products) often drive high quality products out of the market because of the bad reputation they create for other industry products. Consequently, customers require greater assurances to purchase a product in the event it is found after the purchase that the product is a lemon (Akerlof, 1970). The creation of trust represents a particularly significant entry barrier for product innovations that are costly, technologically sophisticated, and whose purchase entails irreversible health or welfare situations for customers. Numerous mechanisms are often established to counteract this quality uncertainty entry barrier, including guarantees, licensing practices, regulatory approvals, as well as endorsements by other trusted institutions. In the case of product guarantees, the risk of quality uncertainty is borne by the seller rather than the buyer.

However, endorsements, licensing, and regulatory approval practices represent industry-wide institutional mechanisms, whose costs are borne both by industry members collectively and by individual firms. They are both products of, and constraints upon, the legitimacy of individual firms engaged in the commercial introduction of a technological innovation. One of the ways in which these firms collectively create and maintain these institutional legitimating devices is through industry councils, technical committees, and trade associations. These industry associations, in turn, approach, educate, and negotiate with other institutions and governmental units to obtain endorsements and develop regulatory procedures.

Few would disagree that all the activities described above in the three subsystems are essential to the emergence and maintenance of an industry. From a macro viewpoint, to understand industry emergence is to know (1) how and when activities in the three functional subsystems are organized, (2) what actors create and perform these functions, and (3) what consequences various arrangements of these functional subsystems have on technological innovation. Seldom can or does a single firm perform all these functions. Thus, from an individual firm viewpoint, three key decisions are made: (1) What functions will the firm perform? (2) What other organizations should the firm link or contract with to have the other functions performed? Consequently, (3) What organizations will the firm compete with on certain functions and cooperate with on others? The next section will discuss these system-level and firm-level process questions.

PART III. SYSTEM AND FIRM PROCESSES IN INDUSTRY EMERGENCE

As these questions suggest, a pragmatic understanding of the implications of the framework presented here requires examining the process of industry emergence at two levels of analysis: (1) the system level looking at the industry as a whole, and the inter-relations between subsystems, and (2) the behavior of individual firms within the industry.

A. Links between Industry Subsystems

The instrumental, resources, and institutional subsystems of an industry are viewed as a loosely-coupled hierarchical system. This loose coupling promotes both flexibility and stability to the structure of an industry. Links between subsystems are only as rich or tight as is necessary to ensure the survival of the system (Aldrich and Whetten, 1981, p. 388). Based on Simon's (1962) architecture of complexity, Aldrich and Whetten discuss how a loosely joined system provides short-run independence of subsystems and long-run dependence only in an aggregate way. The overall social system can be fairly stable, due to the absence of strong ties or links between elements and subsystems, but individual subsystems can be free to adapt quickly to local environmental conditions. Thus, in a complex, heterogeneous, and changing environment, a loosely-joined system is highly adaptive.

> A loosely joined system reaps the benefits of size and specialization of function, and remains flexible enough to cope with a wide range of contingencies. An important feature of hierarchical, loosely joined systems is that although the individual subsystems have wide latitude to adapt to local environments, their adaptation also depends upon constraints placed upon them by other subsystems to which they are vertically and horizontally linked (Aldrich and Whetten, 1981, p. 388).

While there are many linkages between subsystems, we will focus on those links that apply to three pragmatic problems in industry development: (1) resource allocation, (2) the commercial application of basic knowledge, and (3) the emergence of industry standards.

1. *Research Resource Allocation*

A basic practical question confronting both private and public sector research managers is whether they should allocate their research resources to (1) advance knowledge in those fields where scientific and technological opportunities appear most favorable, or (2) for applied R&D in those fields characterized by current or anticipated rapid growth in demand. In other words, will technological innovation proceed more rapidly and efficiently if investments are made to develop basic knowledge in the resource procurement subsystem or if they are allocated to the development of applied knowledge in the instrumental subsystem level? Thirtle and Ruttan (1986) review the long standing debate among economists about whether the source of technical change has been driven primarily by autonomous advances in science and technology (supply or ''technology push'' theories), or driven by economic forces (''demand pull'' theories). This debate intensified in the late 1960s by the HINDSIGHT (1969) study conducted by the Defense Department (Sherwin and Isanson, 1967). The study reported that the significant factors that contributed to the development of 20 major weapons systems were military need rather than disinterested scientific inquiry. These findings were challenged in the TRACES studies by the Illinois Institute of Technology (1968) and the Battelle Research Institute (1973). These studies adopted a much longer time horizon that the 20-year period employed by the HINDSIGHT study. And, not unexpectedly, they found that scientific events were of much greater importance than demand factors as the source of technical change. Thirtle and Ruttan (1986, p. 10), Mowery and Rosenberg (1979), and Dosi (1982) conclude for different reasons that in spite of the extensive research on this question, results to date have been inconclusive.

Our industry framework provides a possible way to reconcile the debate by proposing that both models are likely at play in the emergence of an industry; ''technology push'' is likely to be the dominant force stimulating resource subsystem activities (particularly basic science and technological knowledge), while ''demand pull'' is the major influence directing proprietary instrumental activities.

2. *Commercial Application of Basic Knowledge*

It is generally recognized that resource endowments precede the development of instrumental and institutional subsystem activities of an industry, because basic research (which is the search for a fundamental understanding of natural phenomena) provides the foundation of knowledge that makes possible the commercial birth of a technology (Abernathy, 1978; Rosenberg, 1983; Mowery, 1985). What is less well understood is the process by which a common pool of basic scientific or technological knowledge is transformed into proprietary innovations that can become commercial monopolies. Indeed, as Stobaugh (1985, p. 107) and Mowery (1985) discuss, success at creating a monopoly by commercializing a new technology does not rest on a unique grasp of basic research, but rather on the completion of a long, complex, and highly uncertain journey. This journey consists of an interactive search process involving large amounts of "backing and forthing" between technology and market conditions in efforts to reduce uncertainty.

Unfortunately, very little longitudinal research has been conducted to document these search processes in industry development. Given that technological innovation and industry creation are inherently uncertain, one should expect to observe many trials and errors, as witnessed by multiple divergent and parallel paths over time of competing technological trajectories among firms (Nelson and Winter, 1977). Many of these paths should lead to dead ends, stalemates, and terminations over time. Moreover, an examination of the outcomes or consequences of these paths should lead one to examine if and how learning occurs from previous terminated paths which can provide guidance as to the next paths taken to develop the technological innovation. By examining the outcomes of alternative paths, one could also identify the feasible sets of paths available in the emergence of an industry. While great inter-industry differences should be expected (Mowery, 1985), only by cumulative longitudinal studies of these technological progressions will we come to appreciate the endogenous dynamics of technological innovation and industry creation (Dosi, 1982).

3. *Creation of Industry Standards*

Industry standards pertaining to product testing, reporting, and performance are commonly established to govern product development and commercialization efforts of firms. Some of these standards emerge by cooperative behavior among competing firms when the benefits of standardization outweigh their costs for industry participants. The benefits of standardization include: (1) a reduction in technological uncertainty as standards enable a comparison of different products, (2) greater acceptance of a new product by customers because of baseline quality standards, (3) better policing of exaggerated claims by manufacturers, thereby increasing customer confidence, (4) reduced manufacturing costs, as vendors are better able to supply standardized raw materials and components, and (5) increased speed of regulatory reviews. However, the institution of such standards among competing firms is often a politically-charged process. This is because products based on alternative technological paths may possess fundamentally different product features that appeal to the end user. Consequently, firms will attempt to incorporate those tests, procedures and specifications that favorably measure the attributes of their products into standards that are universally applied across the industry. As Constant (1980) suggests, such standards can inhibit those technological paths which are unable to meet the requirements of standards, even though the product in question may be technologically superior to other available products. For example, the VHS videocassette format was able to displace the technically superior Beta format.

Industry standards can also be mandated by institutional regulatory agencies or trade associations. For instance, the U. S. Food and Drug Administration requires that firms submit a pre-market approval application to seek its approval for the commercial sale of any device. This pre-market approval application covers all three types of standards discussed earlier—testing, reporting and product specifications.

B. Links Among Individual Firms

The social system perspective of an industry emphasizes that any given firm is but one actor, able to perform only a limited set of roles, and dependent on many other actors to accomplish all the functions needed for an industry to emerge and survive. As a consequence, the firm must make strategic choices concerning the kinds of instrumental, resource, and institutional functions in which it will engage, and what other actors it will link with to achieve self-interest and collective objectives. These strategic choices make clear that the firm, and the way it engages in transactions are variables, and that the lines separating the firm from its industrial environment are not sharply drawn, but are fluid and changing frequently over time. These choices and transactions evolve over time, not only as a result of individual firm behavior, but just as importantly by the interdependencies that accumulate among firms across industry subsystems.

Pragmatically, therefore, firm managers and entrepreneurs should be concerned not only with their own immediate instrumental tasks and transaction modes, but also with those of other firms in their resource distribution channel, and with the overall social system. Switching between subsystems and instrumental channels is not inexpensive. Influencing one's own existing channel may be more efficient than switching channels or creating new channels. Also, there is an ongoing tension for each industry participant to organize its own instrumental functions and distribution channels as opposed to contributing to the creation of the industry's resources and institutional subsystems. Although the former may advance the firm's position as a first mover in the short run, the latter provides the infrastructure that ultimately will influence the collective survival of the emerging industry.

There are three significant consequences involved in these decisions by industry participants: (1) cooperative and competitive elements in interfirm relations, (2) multiplexity of relationships, and (3) the collective action dilemma.

1. *Cooperative and Competitive Elements in Relationships*

First, contrary to industrial economists' stress on competitive interfirm relations, the social system perspective emphasizes there are cooperative and competitive elements in each relationship (Van de Ven, Emmett and Koenig, 1974). For example, it is easy to understand that a firm needs to establish cooperative relationships with suppliers, distributors and customers in order to make its own activities meaningful. It is also easy to see that other firms who sell substitutes carry out conflicting activities when they try to take over a customer's patronage. However, as Mattsson (1987) discusses, there are also important elements of conflict in the relations between cooperating firms that have to do with the negotiation and administration of business transactions and adaptation processes. Between instrumental competitors, there are also elements of complementarity, not only when they cooperate to share resources or develop industry institutional functions, but also when they are complementary suppliers to the same customers.

2. *Multiplex Inter-Firm Ties*

Since firms in an emerging industry are often engaged in multiple issues simultaneously, they create a "multiplexity of ties" (Galaskiewicz, 1985, p. 296). Indeed, Aldrich and Whetten (1981, p. 392) point out that it is misleading to think of single relations among most firms in an industry. Common forms of multiple linkages between a given set of firms include: exchanging multiple resources, communications between multiple firm representatives on industry and trade committees, sharing common pools of knowledge, acquiring personnel trained and socialized in a common pool of competence, friendship and kinship ties, and overlapping board memberships. The greater the multiplicity of these relationships the greater the overall stability of an inter-firm relationship. A rupture in one aspect of a relationship does not sever the other ties, and the latter are often used to correct or smooth over the "wounded" link. From an industry perspective, stability through redundance of functions and activities among actors minimizes the negative impact of the loss of services provided by one industry member on the performance of the total system.

Multiple ties among firms emerge over time. Prior relationships and transactions among firms in the pursuit of an industry subsystem activity are remembered, and thereby become the infrastructure upon which subsequent relations are based (Van de Ven and Walker, 1984). Galaskiewicz (1985, p. 299) nicely summarizes some of these temporal dimensions.

> The networks of resource exchange that already existed among organizations are the infrastructure upon which political coalitions are built. In all likelihood, these resource networks were created out of competitive struggle for survival by self-seeking and self-centered actors, who where seeking to minimize their dependencies upon one another. Now these networks are the infrastructure upon which coalitions to achieve collective goals are built. In turn, as political coalitions become institutionalized, they impinge on the struggle for dominance in the resource procurement/allocation arena.

An appreciation of the temporal dimension of inter-firm relationships also provides important insight on how competitors emerge in an industry. Generally, the literature tends to assume that competitors are profit-seeking entrepreneurs who somehow recognize and seize commercial opportunities by entering lucrative markets. Based on their longitudinal study of the emergence of the cochlear implant industry, Garud and Van de Ven (1989) provide quite a different explanation for how industrial competitors emerged. *Aborted efforts at establishing cooperative relationships turn out to become competitive relationships.*

> In two instances, the efforts of the first mover to initiate cooperative relationships or joint ventures with other research clinics failed, leading to the birth of the firm's competitors. Initial negotiations of possible relationships with a foreign university and a domestic university did not materialize. Otological scientists and clinicians in each of these two universities subsequently entered into licensing arrangements with two other firms (one a new company startup, the other a subsidiary of a large manufacturer), who now two years later are the first-mover's major competitors (Garud and Van de Ven, 1988, p. 14).

In this process, key "linking-pin" organizations emerge that have extensive and overlapping ties to different subsystems of the industry, and play the key role of integrating the system. Because they have ties to more than one subsystem, these linking-pin firms are the nodes through which a network is loosely joined (Aldrich and Whetten, 1981). They serve as communication channels between industry participants, and link third parties by transferring resources, information, or specialties within and outside of the industry. By being linked into multiple subsystem functions in the industry, these linking pin organizations accumulate a broad base of power for ascending to a dominant position in the industry, which permit them to survive at the expense of peripheral industry participants.

But these linking-pin organizations also experience the greatest conflicts of interest in the industry, since they tend also to have the least amount of autonomy and ability to capture significant proprietary advantages. This is because their dominance serves as a model that is imitated by others and diffused throughout the industry. Key linking-pin firms, who are often the first to introduce products into the market, must also bear significant "first mover" burdens which permit free-riding by other industry participants. But in return for these burdens, the first mover gains the greatest degree of freedom to shape industry rules and product perception in such ways that benefit it the most (Porter, 1985).

3. *Paradox of Cooperation and Competition*

Inherent in all the above relationships among firms engaged in an emerging industry is the paradox of cooperation and competition. Each firm competes to establish its distinctive position in the industry; at the same time, firms must cooperate to establish the industry infrastructure. Olson (1965, p. 10) introduces the paradox:

> If firms in an industry are maximizing profits, the profits for the industry as a whole will be less than they might otherwise be. Almost everyone would agree that this theoretical conclusion fits the facts for markets characterized by pure competition. The important point is that this is true because, though all the firms have a common interest in a higher price for the industry's product, it is in the interest of each firm that the other firms pay the cost—in terms of the necessary reduction in output—needed to obtain a higher price."

Another example that pertains to the institutional subsystem level is that it clearly benefits all firms to cooperate to set up industry standards. However, in doing so, each firm will try to ensure that the industry standards which suit it best get institutionalized.

One of the major reasons that has been offered for the origin of industry regulation is that it is an institutional means to address these collective action dilemmas, in which individual firms do not voluntarily act in a designated way to achieve benefits for all industry participants (Mitnick, 1980, pp. 164–165). Institutional ways to guarantee such action must be devised to provide the benefits. Otherwise, individual self-interest may lead some members to "free ride" on whatever group benefits may have been obtained by others.

CONCLUDING DISCUSSION

The purpose of this chapter has been to understand how new industries emerge. We found that the traditional industrial economic view of an industry and its natural selection theory of price competition were not adequate for explaining technological innovation and industry emergence for two reasons. First, the traditional definition of an industry needs expansion not only to include the group of firms competing to produce similar or substitute products, but also include many other actors and institutions that perform key roles necessary to transform a technological innovation into a commercially viable line of products delivered to customers. Although these other actors and institutions have heretofore been treated as "externalities" (Porter, 1980), they are critical units to incorporate directly in an explanation of industry emergence. Second, we reviewed arguments why the natural selection theory of price competition cannot explain industry emergence, and why the punctuated equilibrium model of change is useful to describe the overall life cycle of an industry. However, the latter does not adequately explain the process of "punctuation" itself.

As an alternative, we proposed that an industry be viewed as a social system consisting of three loosely-coupled hierarchical subsystems: Instrumental, resource procurement, and institutional subsystems. We also suggested that the process of industry emergence is best described by an accumulation theory of change that includes overlapping periods of initiation, startup and takeoff. This framework is broader than most treatments of industries, and combines relevant perspectives from industrial economics, marketing channels, and organizational sociology. We also examined some of the practical implications of the proposed framework and theory of change for studying the processes of industry emergence, both from a macro perspective of the industry as a whole, as well as from a micro perspective of individual firms constituting the industry.

In conclusion, we will discuss some of the limitations and strengths of the proposed framework. After all, frameworks act as "lenses," filtering what we see, and how we interpret phenomena (Allison, 1971). Thus, the test of a good framework is its ability to offer insights about the phenomena under observation by separating the important from the unimportant issues and by helping define the critical subproblems to be dealt with (Rumelt, 1979, p. 199). At the same time, any framework has its weaknesses. We acknowledge that this framework too has its weaknesses. For instance, incorporating three subsystems as a part of the industry social system considerably increases the complexity of one's perspective. This increased complexity could act as a significant barrier preventing policy makers from coming to a decision about a clear course of action. This could be either because of their inability to gather the necessary data, or because of their inability to make the analytic connections between the subsystems, thereby resulting in "analysis paralysis." However, we will conclude by pointing to three major benefits of the framework.

1. Holistic Perspective

It is all too easy to study any one aspect of an industry without regard to others. For instance, the five forces model (Porter, 1980) focuses upon the transactional environment of any given firm while the resource channel approach of marketing (Stern and El-Ansery, 1982) focuses on the linear flow of goods and services to the customer. While important insights can be gained by focusing on parts of the whole, there is a danger of developing "tunnel vision," where the solutions to one set of problems result in the creation of other problems elsewhere. The proposed social system perspective not only safeguards against taking too parochial a view, but also permits us to synthesize different bodies of knowledge and subfields to gain a deeper appreciation of the process of technological innovation and industry emergence. Technological innovation encompasses a wide array of problems and issues that can be addressed with the social system framework—such as the development of institutional governance structures, the creation of basic resources to support an industry, questions of "technology push" and "demand pull," patterns of cooperative and competitive relationships among firms, firm-specific problems of product development and business creation, and overall industry structure.

2. Part-Whole Relationships

The social system framework helps highlight important micro-macro linkages. For instance, while the five forces model implicitly suggests that businesses should focus upon developing distinctive competences to gain competitive advantage over others, the social system framework suggests that firms should be as much concerned with promoting cooperative arrangements for creating the industry's infrastructure as they are with gaining instrumental first-mover advantages during its emergence. It also brings into sharper focus some of the dilemmas that firms face—whether to use their resources to develop instrumental level activities or to develop the industry infrastructure that could benefit all.

3. Dynamic Interplay Between Industry Subsystems

Finally, the social system perspective facilitates an appreciation of the longitudinal relationships among components of an industry. The framework allows us to examine the dynamic interplay in the progression of the different subsystems over time, as we have shown elsewhere in the emergence of the cochlear implant industry (Garud and Van de Ven, 1988). This framework also enables one to examine the different paths by which an industry emerges over time, and how these paths affect the progression of other paths and eventually the emergence of the entire industry. This in turn makes it possible to understand the time duration for the emergence of a self-sustained industry by stringing together the time required for the occurrence of necessary events that crisscross between different subsystems.

ACKNOWLEDGMENTS

We gratefully acknowledge useful comments on earlier drafts of this chapter from Harold Angle, Graham Astley, Robert Burgelman, Bala Chakravarthy, Joseph Galaskiewicz, Edward Layton, John Mauriel, Scott Poole, Michael Rappa, Peter Ring, Richard Rosenbloom, Vernon Ruttan and other colleagues involved in the Minnesota Innovation Research Program at the University of Minnesota.

Support for this research program has been provided by a grant to the Strategic Management Research Center at the University of Minnesota from the Program on Organization Effectiveness, Office of Naval Research (code 4420E), under contract No. N00014-84-K-0016.

NOTES

1. Of course, hierarchy in an industry system is a matter of degree, and some industry systems may be only minimally if at all hierarchical. Hierarchy is often a consequence of institutional constraints imposed by political and governmental regulatory bodies. Hierarchy also emerges in relationships with key linking-pin organizations who either become dominant industry leaders or control access to critical resources (money, competence, technology) needed by other firms in the industry.

2. Parson's functional differentiation of subsystems has been a common theme underlying structural descriptions of complex organizations (Thompson, 1967; Pfeffer and Salancik, 1978), corporations (Porter, 1985) and interorganizational networks (Van de Ven, Emmett and Koenig, 1974; Aldrich and Whetten, 1981).

REFERENCES

Abernathy, W. J., 1978. *The productivity dilemma: Roadblock to innovations in the automobile industry.* Baltimore: John Hopkins University Press.

Abernathy, W. J. and K. B. Clark, 1983. "*Innovation: mapping the winds of creative destruction.*" Cambridge, Mass.: Harvard Business School Working Paper 84–32 (July 25).

Akerlof, G. A., 1970. "The market for 'lemons': Quality, uncertainty and the market mechanism." *Quarterly Journal of Economics,* 84, pp. 488–500.

Alchian, A. A. and H. Demsetz, 1972. "Production, Information Costs, and Economic Organization." *American Economic Review,* 65, 5, pp. 777–795.

Aldrich, H., and Whetten D., 1981. Organization-sets, action sets, and networks: Making the most of simplicity. In P. C. Nystrom & W. H. Starbuck (Eds.), *Handbook of Organizational Design,* pp. 385–407. Oxford: Oxford University Press.

Allen, R. C., 1983. "Collective Invention." *Journal of Economic Behavior and Organization,* 4, 1 (March), pp. 1–24.

Allison, G. T. 1971. *The essence of decision: Explaining the Cuban missile crisis.* Boston: Little Brown, & Co.

Arrow, K. J., 1962. "Economic Welfare and the Allocation of Resources for Innovative Activity," in R. R. Nelson (ed.), *The Rate and Direction of Inventive Activity,* Princeton, New Jersey: Princeton University Press.

Astley, W. G. 1985. "The two ecologies: Population and community perspectives on organizational evolution." *Administrative Sciences Quarterly,* Vol. 30, pp. 224–241.

Astley, W. G., and Van de Ven A. H., 1983. "Central perspectives and debates in organization theory." *Administrative Sciences Quarterly,* Vol. 28, pp. 245–273.

Battelle Research Institute, 1973. *Interaction of science and technology in the innovation process: some case studies.* Columbus, Ohio: Battelle Research Institute.

Commons, J. R., 1950. *The economics of collective action.* Madison, Wisconsin: University of Wisconsin Press.

Constant, E. W., 1980. *The origins of the turbojet revolution.* Baltimore: The John Hopkins University Press.
DiMaggio, P. J., 1982. "*The structure of organizational fields: An analytic approach and policy implications.*" Paper presented for SUNY-Albany Conference on Organizational Theory and Public Policy.
DiMaggio, P. J. and Powell, W., 1983. "The iron cage revisited: Institutional isomorphism and collective rationality in organizational fields." *American Sociological Review,* Vol. 48, pp. 147–161.
Dosi, G. 1982. "Technological paradigms and technological trajectories." *Research Policy.* Vol. 11, pp. 147–162.
Durkheim, E., 1933. *The division of labor in society.* New York: MacMillan.
Etzioni, A., 1963. "The epigenesis of political communities at the international level." *American Journal of Sociology,* 58, 407–421.
Galaskiewicz, J. 1985. "Interorganizational relations." *Annual Review of Sociology,* Vol. 11, pp. 281–304.
Garud, R., and Van de Ven, A. H., 1989. Innovation and the emergence of industries. In A. H. Van de Ven, H. Angle, and M. S. Poole (Eds.), *Research on the Management of Innovation.* Cambridge: Ballinger. Forthcoming.
Gould, S. J. and N. Eldredge, 1977. "Punctuated equilibria: the tempo and mode of evolution reconsidered." *Paleobiology,* 3, pp. 115–151.
Granovetter, M., 1985. "Economic action and social structures: The problems of embeddedness." *American Journal of Sociology,* Vol. 91, No. 3, pp. 481–510.
Hannan, M. and J. Freeman, 1977. "The population ecology of organizations." *American Journal of Sociology,* 82, pp. 929–964.
HINDSIGHT, 1969. *Project Hindsight: Final Report,* Washington, D.C., Office of the Director of Defense Research and Engineering.
Hirsch, P. M. 1975. "Organizational effectiveness and the institutional environment." *Administrative Science Quarterly,* Vol. 20, pp. 327–344.
Illinois Institute of Technology, 1968. *Technology in retrospect and critical events in sciences (TRACES).* Mimeographed, Illinois Institute of Technology, Chicago.
Kamien, M. I., and N. L. Schwartz, 1982. *Market structure and innovation,* Cambridge: Cambridge University Press.
Mansfield, E., 1985. "How rapidly does new industrial technology leak out?" *Journal of Industrial Economics* 34, 2 (December), pp. 217–223.
Mattsson, L. G., 1987. "Management of strategic change in a "markets-as-networks" perspective," in A. Pettigrew (ed.), *The Strategic Management of Change.* London: Basil Blackwell.
McKelvey, B., 1982. *Organizational systematics: taxonomy, evolution, classification,* Berkeley: University of California Press.
Metcalfe, J. S., and L. Soete, 1983. *Notes on the evolution of technology and international competition.* Paper presented at the workshop on Science and Technology Policy, University of Manchester, (April).
Meyer, J. W. and Rowan, B., 1977. "Institutionalized organizations: Formal structure as myth and ceremony." *American Journal of Sociology,* Vol. 83, No. 2, pp. 340–363.
Miles, R. and Cameron, K. S., 1982. *Coffin nails and corporate strategies,* Englewood Cliffs: Prentice-Hall Inc.

Mitnick, B. M., 1980. *The Political Economy of Regulation: Creating, Designing, and Removing Regulatory Forms,* New York: Columbia University Press.

Mowery, D. C., 1985. *Market structure and innovation: A critical survey.* Paper presented at the conference on "New technology as organizational innovation" at the Netherlands Institute for Advanced Studies in Humanities, Wassenaar.

Mowery, D. C., and Rosenberg, N., 1979. The influence of market demand upon innovation: A critical review of some recent empirical studies. *Research Policy,* April.

Mueller, W. F., 1962. "The Origins of the Basic Inventions Underlying DuPont's Major Product and Process Innovations, 1920–1950," in R. R. Nelson (ed.), *The Rate and Direction of Inventive Activity,* Princeton, New Jersey: Princeton University Press.

Nelson, R. N., 1959. "The Simple Economics of Basic Scientific Research." *Journal of Political Economy,* 67, pp. 297–306.

Nelson, R. N., and Winter, S. G., 1977. "In search of a useful theory of innovation." *Research Policy.* Vol. 6, pp. 36–76.

Nelson, R. N., and Winter, S. G., 1982. *An Evolutionary Theory of Economic Change.* Cambridge: Harvard University Press.

Olson, M., 1965. *The logic of collective action: Public goods and the theory of groups.* Cambridge: Harvard University Press.

Parsons, T., 1964. *The social system.* New York: Free Press.

Pfeffer, J., and Salancik, G., 1978. *The external control of organizations.* New York: Harper & Row.

Piore, M. J., and C. F. Sabel, 1984. *The second industrial divide: possibilities for prosperity,* New York: Basic Books.

Porter, M. E., 1980. *Competitive strategy: Techniques for analyzing industries and competitors.* New York: The Free Press.

Porter, M. E., 1985. *Competitive advantage: Creating and sustaining superior performance.* New York: The Free Press.

Rappa, M., 1987. *The structure of technological revolutions: An empirical study of the development of III-V compound semiconductor technology.* Unpublished Dissertation, University of Minnesota, Carlson School of Management, Minneapolis, Minnesota.

Rosenberg, N., 1983. *Inside the black box: Technology and economics.* Cambridge: Cambridge University Press.

Rosenbloom, R. S., 1966. "Product Innovation in a Scientific Age." *New Ideas for Successful Marketing.* Chicago: Proceedings of the 1966 World Congress, American Marketing Association, chapter 23.

Rosenbloom, R. S., 1985. "Managing technology for the longer term: a managerial perspective," chapter 7 in K. B. Clark, R. Hayes, and C. Lorenz (eds.), *The Uneasy Alliance: Managing the Productivity Technology Dilemma,* Cambridge, Mass.: Harvard Business School, pp. 297–317.

Rumelt, P. 1979. "Evaluation of strategy: Theory and models," in Schendel, D. E., and C. W. Hofer (eds.), *Strategic management: A new view of business policy and planning.* Boston: Little, Brown and Company.

Sahal, D., 1981. *Patterns of technological innovation,* Reading, MA: Addison-Wesley.

Schumpeter, J. A., 1975. *Capitalism, socialism, and democracy,* New York: Harper and Row.

Sherwin, C. W., and R. S. Isenson, 1967. "Project Hindsight: a defense department study of the utility of research." *aScience,* 156, pp. 1571–1577.

Simon, H. A., 1962. "The architecture of complexity," *Proceedings of the American Philosophical Society,* 106, pp. 467–482.

Smith, A. 1937 (1776), *The wealth of nations,* New York: Modern Library.

Stern, N., and El-Ansery, A. I., 1982. *Marketing Channels.* New Jersey: Prentice Hall.

Stigler, G. J., 1957. "Perfect competition, historically contemplated." *Journal of Political Economy,* 65, pp. 1–16.

Stobaugh, R., 1985. "Creating a monopoly: Product innovation in petrochemicals," in R. Rosenbloom (ed.), *Research on technological innovation, management and policy.* Vol. 2, pp. 81–112. New York: JAI Press.

Thirtle, C. G., and Ruttan, V. W., 1986. *The role of demand and supply in the generation and diffusion of technical change.* Bulletin No. 86-5. University of Minnesota Economic development center, Minnesota.

Thompson, J. D., 1967. *Organizations in action,* New York: McGraw-Hill.

Utterback, J. M., 1974. "Innovation in Industry and the Diffusion of Technology," *Science,* Vol. 183, pp. 620–626.

Van de Ven, A. H., and Astley W. G., 1981. "Mapping the field to create a dynamic perspective on organization design and behavior," in A. Van de Ven and W. Joyce (eds.), *Perspectives on organization design and behavior,* pp. 427–468. New York: John Wiley & Sons.

Van de Ven, A. H., D. Emmett, and R. Koenig, 1974. "Alternative Frameworks for Interorganizational Analysis," *Organization and Administrative Sciences,* 5, 1, pp. 113–129.

Van de Ven, A. H., and M. S. Poole, 1987. "Paradoxical requirements for a theory of organizational change." Chapter 1 in R. Quinn and K. Cameron (eds.), *Paradox and transformation: toward a theory of change in organization and management.* Cambridge, Mass.: Ballinger.

Van de Ven, A. H., and G. Walker, 1984. "The dynamics of interorganizational coordination." *Administrative Science Quarterly,* 29, pp. 598–621.

von Hippel, E. 1986. "Cooperation between rivals: informal know-how trading." Cambridge, Mass.: MIT Sloan School Working Paper #1759–86.

Warren, R., 1971. *Truth, Love and Social Change and other Essays on Community Change,* Chicago: Rand McNally.

Williamson, O. E., 1975. *Markets and hierarchies.* New York: Free Press.

Williamson, O. E., 1979. "Transaction-cost economics: the governance of contractual relations," *Journal of Law and Economics,* 22, pp. 233–261.

Yin, L., and Segerson, P. E., 1986. Cochlear implants: Overview of safety and effectiveness. *Otolaryngologic Clinics of North America.* Vol. 19, no. 2, pp. 423–433.

COLLABORATIVE ARRANGEMENTS AND GLOBAL TECHNOLOGY STRATEGY:

SOME EVIDENCE FROM THE TELECOMMUNICATIONS EQUIPMENT INDUSTRY

Gary Pisano and David J. Teece

I. INTRODUCTION

The organization of the technological innovation process is changing worldwide, and particularly in the United States. These changes encompass both the way that research is organized and the way that new technology is commercialized. The traditional linear approach—R&D, prototyping, manufacturing startup, marketing and distribution all in-house—is giving way, particularly in some new industries, to less vertically integrated structures, usually involving collaboration with other industry

Research on Technological Innovation, Management and Policy
Volume 4, pages 227–256

ISBN: 0-89232-798-7

participants. Indeed, collaboration among unaffiliated enterprises for manufacturing and/or distribution has become the norm in some industries.

In this chapter we attempt to build a framework to explain this phenomenon. We hope this framework will assist in the design of appropriate technology strategies. In particular, we argue that technology strategy is no longer a matter of setting the R&D budget and selecting promising projects. Technology strategy also involves making key organizational decisions with respect to commercialization. These decisions relate in particular to whether such endeavors should be organized in-house or should involve the participation of unaffiliated firms, including potential rivals and competitors. They also involve decisions related to the governance of collaborative relationships. The framework presented in this paper suggests that different inter-firm governance structures will be appropriate for different types of collaborative activities. While there are many possible forms of inter-firm governance, in this paper we distinguish between relationships involving equity participation and those governed strictly by contracts. Section VI investigates the choice between equity and non-equity modes of governance in the context of technological collaborations in the telecommunications equipment industry.

Of course, interfirm collaborations are not new. They are common practice both domestically and internationally. What is new is the frequency with which collaborations of this kind are occurring in the development and commercialization of new products and processes. In order to give some precision to the discussion, we first provide a definition of collaboration as well as some examples.

We define collaborative agreements in the context of innovation as any interorganizational (firm, university, government lab, etc.) agreements, with or without equity, that involve the bilateral, multiparty, or unilateral contributions or exchange of assets or their services in a market. We are deeply concerned with such basic organizational issues as whether markets or hierarchies should be used to organize economic activity.

Collaboration under the definition just provided is not confined to the innovation process. Any joint venture, such as the famous ARAMCO and CALTEX joint ventures in the petroleum business, or COMALCO in aluminum, would qualify. Accordingly, we hereafter confine our analysis to collaborative arrangements that involve the development and/or commercialization of new products and processes. We have in mind the kind of collaboration witnessed between Lilly and Genentech for the development and commercialization of Humulin, between Merck and Chiron for Hepatitis B, between AT&T and Olivetti for the AT&T PC, and between McDonnell Douglas and Northrup for the F-18 jet fighter. All of these arrangements involve new products and processes. Human capital, rather than physical capital, is what is being shared.

II. THE ORGANIZATIONAL DIMENSIONS OF TECHNOLOGY STRATEGY

There are many elements of technology strategy. Firms must select projects, allocate R&D budgets, decide how much they are going to lay on those bets, and organize themselves in order to give those projects the best chance of generating positive commercial outcomes. The first two of these dimensions is captured on the vertical (y) axis in Figure 1. The third is represented in the other 2 axes (x and z). The more traditional model for firms has been vertical integration (shaded area in Figure 1), as illustrated by IBM's mainframe computer business. Increasingly, however, less integrated approaches are appearing. For example, Worlds of Wonder, Inc., a "producer" of high technology childrens' toys, contracts out for all of its manufacturing and some of its design.

The organizational question can thus be partitioned into two sets of choices. The first one relates to whether the established firm ought to rely on its own internal R&D capabilities or whether it should source technology externally.[1] The second relates to the organizational approach the firm should adopt for commercializing the technology. Should it attempt all of the relevant manufacturing and marketing in-house, or should it seek others with whom it can contract for these services? The firm may choose to perform all of the key functions in-house (and be what we call a "classical firm"), or it may choose to collaborate with other more qualified firms to ensure the product is presented to the market in the most attractive way (what we call a "network firm").[2]

In essence, organizational choices we have identified raise the question of whether the firm should use market modes (contracting/venturing with nonaffiliated firms), internal nonmarket modes, or mixed modes (contracting for some functions and performing others in-house). These organizational questions are not typically considered part of technology strategy. Technology strategy was once thought to consist of determining the level of R&D expenditures and the projects to which those resources ought to be allocated.

There are a number of factors that suggest that a much broader concept of technology strategy is now warranted:[3]

1. The frequency of technological discontinuities or technology paradigm shifts seems to have increased. When technological develop-

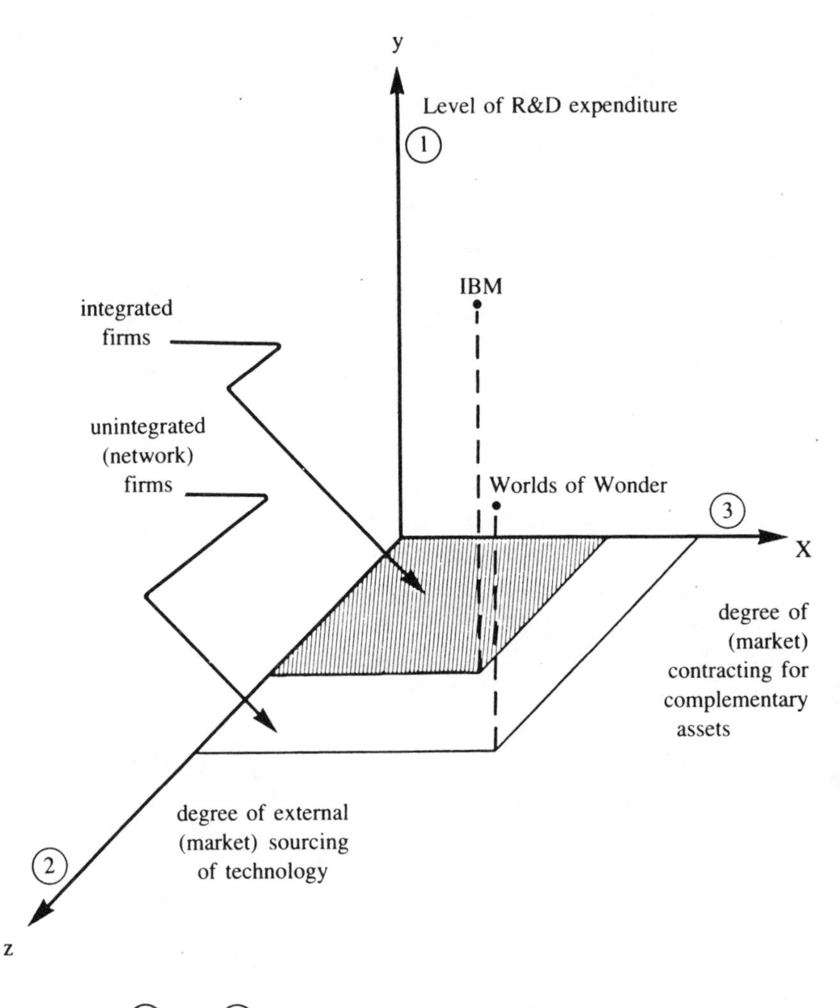

Figure 1. Organizational and Resource Dimensions of Technology Strategy

ment takes a new trajectory, the direction of technical development is no longer cumulative and self-generating. Development requires repeated reference to the technical and commercial environment external to the firm. In short, the logic of previous technical advance is broken; and the capabilities that the firm possesses in-house may no longer suffice. Technological discontinuities have been a feature of technological advance since time immemorial, but according to one source they are on the increase (Foster, 1986).

2. The costs of innovation have increased markedly, and the ability of a single firm to "go it alone," particularly with respect to large systems (e.g., the Boeing 767 or the Airbus A300), may have declined. In short, even setting aside matters of risk, the financial requirements may strain even giant enterprises.
3. The sources of innovation have become more diffused internationally; thus, the probability that any one firm, even if it is multinational, could command all the relevant expertise for a particular project is declining. Certainly, the technological dominance of American firms in several industries is being challenged.
4. The speed with which new technologies must be commercialized has increased to the point where few firms have the time to assemble all of the requisite capabilities in-house. In part, this is because of more rapid technological change. It is also because of rapid imitation. Accordingly, there are cases where collaboration with other firms that already have the requisite capabilities is appealing.
5. For large firms, the incentive properties of small firms—and, in particular, their ability to reward innovators handsomely in ways that large established firms find difficult to replicate—favor the pursuit of technical opportunities externally.

All of this suggests that simple-minded representations of innovation as ideally suited to the large-scale enterprise or to the small firm need to yield to a more eclectic view of innovation and its organizational and strategic requirements. In particular, we indicate that the key strategic/organizational issues for innovating firms to consider are the extent to which technology will be sourced internally, rather than externally, and the degree of involvement that other firms will be offered in the commercialization process.

III. FORCES DETERMINING TECHNOLOGY SOURCING STRATEGIES

Whether innovating firms ought to source technology internally or externally, we submit, depends on the interrelationships between three key sets of factors, each of which will be explored in turn:

1. The organizational location of the sources of invention/know-how
2. The ease of appropriability, i.e., whether the sponsoring firm is well positioned to capture the benefits from the research activity in question
3. The facility with which contracts for the purchase or sale of the technology in question can be written, executed, and enforced. (This can be referred to as the transaction cost.)

A. Sources of Know-how

Winter (1984) argues that innovation involves mixing public know-how, proprietary know-how external to the firm (imitation), and internal know-how. When innovation is driven primarily by internal know-how and capabilities, a firm's ability to exploit technical opportunities is constrained primarily by its accumulated stock of proprietary know-how, its organizational and learning skills, and its experience in the relevant activities. These assets take time to build and are a function of past activities in both research and production. Capabilities relevant to a particular technological paradigm over time become imbedded in its research routines (Nelson and Winter, 1982).

The skills, know-how, and experience necessary to innovate in one design paradigm, however, are usually quite different from those required in another. Thus, if a firm's established technological trajectory is particularly rich, or if it is able to lead the shift in an industry from one design/technological regime to another, then that firm may be able to continue relying on internal capabilities to generate relevant know-how. However, shifts in technological regimes are often propelled by firms that do not have the deepest skills in the established paradigm. Thus, when the transistor replaced the vacuum tube in the mid-1950s, vacuum-tube manufacturers were not the pioneers. Indeed, Sylvania kept pouring money into increasingly sophisticated vacuum-tube designs until 1968! Rather, an entirely new set of producers emerged to develop and produce transistors (Malerba, 1985).

Hence, a shift in technological paradigms (a technological discontinuity) is likely to cause a shift in the locus of the most productive R&D efforts in a direction away from the incumbents. Incumbent firms will thus have to consider acquiring this technology. They may be able to do so through naked imitation; however, as we shall see, if the innovation in question is protected, the technology may need to be purchased externally.

Note that public institutions—universities and government laboratories—may be important sources of new technology, particularly in the early stages of an industry. Inasmuch as such establishments are unable or unwilling as a matter of policy to engage in commercialization activities, the requirement and the opportunity for collaboration with established firms are provided. In these circumstances, firms—both incumbents and new business ventures—are forced to seek technology externally.

B. Appropriability Issues

If new technology can be sourced externally or developed internally, then the choice of mode will likely depend upon appropriability and related transactions costs issues. If certain R&D resources allocated either externally or internally can be expected to produce equally beneficial outcomes, appropriability concerns are likely to favor internal procurement for at least two reasons. One is the cumulative nature of learning, and its particular location. The procuring firm, should it "contract out," is likely to deny itself important learning opportunities. If "one shot" improvements along a particular technological trajectory, as with research to meet a particular fixed regulatory standard, are all that is contemplated, then permitting the developer to benefit from learning may not present a problem. Generally, however, future advances are contemplated, and if these can profitably build upon earlier R&D activity, internalizing the activity will be necessary. Even though the developer may pass on the benefits of past learning acquired under previous R&D contracts with the procurer, there are circumstances under which this may not occur. In-house research guards against these contingencies. A second and related reason is that the unaffiliated developer of new products and processes is generally free to contract with other procurers. This can result in the leakage of technology, developed on one company's R&D dollar, to another. Internalization forces an exclusive contract, and avoids this spillover occuring through the R&D contracting process. It may still of course occur in other ways, as when R&D personnel switch employment.

The desirability of an external sourcing approach increases if the sources of relevant technology are external to the firm.[4] However, even if new technology emerges elsewhere, a firm's eagerness to engage in a commercial transaction to secure the technology depends on the viability and necessity of a contract for the technology in question. If the technology is of a kind for which intellectual-property law affords no protection and if copying is easy, then the acquiring firm need only imitate; and no commercial transaction results. If, on the other hand, the technology at issue is valuable and is protected by patents, trade secrets, and other legal structures, or is simply difficult to copy, then some kind of formal purchase contract and/or technology-transfer agreement will be called for.

While there are many exceptions, it is generally the case that very little know-how can be shielded effectively through patent and trade-secret protection alone. One major exception is chemical-based technologies, where patent protection, due to the nature of the technology, is intrinsically stronger. Xerography (Xerox) and instant photography (Polaroid) are other major exceptions. Patent protection is generally much weaker in machine and process equipment technologies because the nature of the technology makes it vulnerable to reverse engineering (see Levin et al., 1984).

If property rights are very strong, the innovator's reluctance to license is often overcome because the possibility of extracting an economic return comparable to that which could be obtained internally is increased. Conversely, when intellectual-property protection is weak, new technology, if it is developed at all, will be developed internally for internal use. According to Von Hippel (1982), the dominance of equipment users as innovators can be explained by this relative appropriability advantage. In short, strong patent protection allows the innovator to market its product or process innovation without exposing it to risks of imitation. In contrast, a tighter link, possibly even vertical integration, between sources and users of technology, is required when patent protection is weak. Integration, of course, enables trade-secret protection to shield the technology from would-be scrutinizers of the technology. The corollary is that users will have to come to terms with the sources of technology when they are external. If the appropriability regime is weak, they may simply be able to imitate. If it is strong, then imitation is less viable; and some kind of licensing arrangement may have to be sought. The viability of this depends, in turn, on transactions-cost considerations, which we now examine.

C. Transactions Costs

Transactions costs considerations lay behind the appropriability issues previously discussed. Transactions costs relate to the ease with which contracts for the purchase or sale of a commodity, in this case technology, can be written, executed, and enforced without leading to unexpected outcomes that impose large costs on one or both parties (Williamson, 1975, 1985; Teece, 1981). In the case of technology, license agreements are risky if one or both parties must make highly dedicated investments whose value depends on the other party's performing as anticipated.

The biggest transactional risks for the seller are associated with the buyer's using the technology in ways not anticipated by the contract, or which while anticipated cannot be easily prevented. These risks are usually ameliorated if the technology has good protection under relevant intellectual-property law. The biggest risks for the buyer stem from the fact that the technology may not perform at expected levels. The problem stems from the fundamental paradox of information: one often does not know what one has purchased until after the fact (Arrow, 1962). In short, a buyer must typically engage in a transaction in which he has incomplete information about the commodity being purchased (Teece, 1981).

Delivery is another problem. Technology must be transferred from seller to buyer for the transaction to be complete. This can be costly. Unless the technology is highly codified, transfer is likely to involve the transfer of technical personnel; and, depending on the complexities of the technology and the way in which the transfer is managed, the success of the transfer is uncertain (Teece, 1977). In short, the viability of a market relationship involving collaboration will be drawn in part by the transactions-cost conditions that will characterize the contract. It ought to be evident that high transactions costs will block an arrangement even when it would be warranted on other grounds. Such a condition is commonly referred to as "market failure." Market failures are nonevents. They cause deals to be avoided because it is not possible to formulate and/or enforce a mutually acceptable arrangement between buyer and seller. Figure 2 summarizes the implications for know-how procurement strategies that follow from the interactions among the loci of innovation, appropriability issues, and transactions costs.

IV. THE EXTERNAL PROCUREMENT OF KNOW-HOW

The external procurement of know-how is likely to be imperative for incumbent firms that, for one reason or another, find (1) that they are no longer capable of productively researching opportunities internally, and (2) that there are significant costs, particularly from having investment in traditional products stranded, associated with being closed out of the next

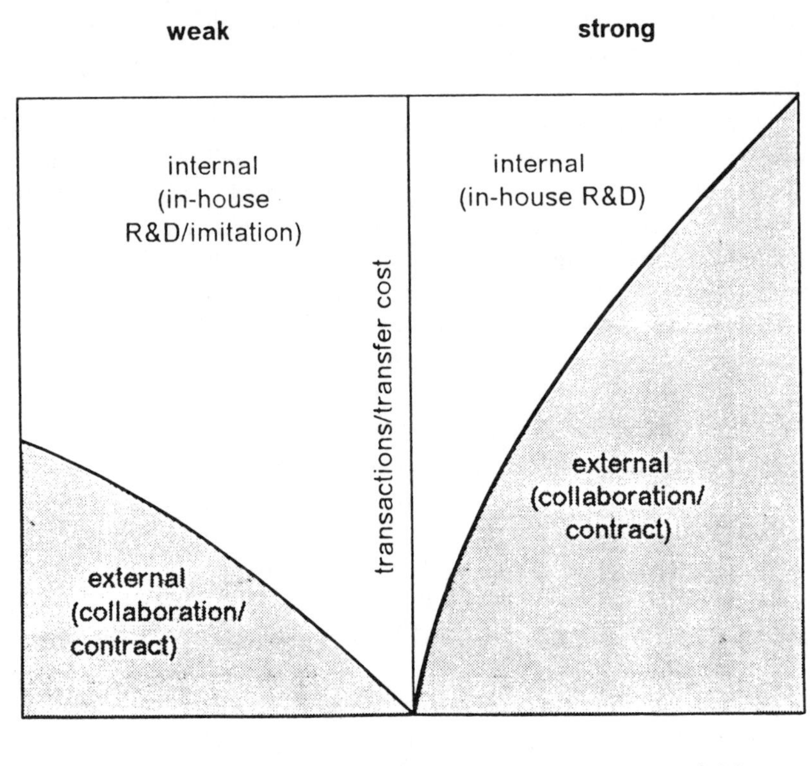

Figure 2. Know-how Procurement Strategy (Buyer's Perspective)

round of significant innovation. The former set of factors might be thought of as pro-active factors favoring external sourcing, while the second kind are reactive or defensive. Each is discussed in turn.

A. Pro-active Considerations

The external procurement of know-how by contract is likely to be the selected sourcing strategy when the centers of excellence in the relevant know-how are external to the firm, appropriability is tight, and transactional difficulties are manageable. The external sources may be suppliers, as in agriculture and housing, users, as in scientific instruments, or competitors, as in biotechnology, or not-for-profit organizations, such as universities and government laboratories.

The locus of innovation is most likely to lie external to incumbent firms when there is a technological discontinuity, or a paradigm shift as it might more meaningfully be called. Of course, what is external to one group of firms may be internal to another.

There are a number of modes by which technology lying external to the firm can be acquired. If it is easy to copy because it lacks intellectual-property protection and can be reverse-engineered at low cost, as with some microprocessors, then imitation is often a viable acquisition strategy. When the technology is legally protected, is hard to copy, and the innovator is willing to sell, then a number of possible contractual relationships are possible. They include licensing, contract R&D, R&D joint ventures, and bilateral collaborative arrangements.

Licensing is the most familiar of these approaches. A firm possessing valuable know-how that is protected can contract to let others use the technology in question. A license agreement will often be accompanied by a know-how agreement under which the owner of the intellectual property in question will contract to assist the buyer in developing a comprehensive understanding of the technology in question.

Contract R&D is also an important mode, though it is also fraught with hazards. When a buyer commissions R&D work to be performed under contract, it is usually in recognition of the fact that the provider of the R&D services is better positioned to generate a desirable output from R&D than is the buyer itself. Unless the technology to be developed can be specified with great precision, and the costs of the requisite development activities can be gauged with considerable accuracy, however, contracting to develop technology using fixed-price contracts is not easy as it is difficult to specify and cost-out the object of the development activity at issue. Modest technological endeavors can be arranged this way more satisfactorily than can ambitious ones (which can typically be organized externally only by cost plus contracts, but are exposed to obvious incentive hazards).

R&D joint ventures make sense as external procurement mechanisms when the other party can bring certain capabilities to the venture that the collaborating party does not possess. Other properties of joint ventures are that they reduce risk when project costs are high; and in the R&D area, they may reduce duplication without necessarily reducing variety.

An inherent flaw of capitalist-market economics is that they often cause patent races and other forms of socially wasteful R&D duplication. There may not be a better system for promoting innovation than capitalism, and gross inefficiencies associated with duplication can be reduced by joint ventures. Research consortia, such as the Microelectronics and Computer Corporation (MCC) formed by a group of computer companies, are an example.

Another collaborative mechanism involves bilateral exchanges of know-how and other assets, as with cross-licensing, patent-pooling, and, more recently, technology transfers, in return for some other nonpecuniary commercial favor, such as access to distribution facilities. These services are often difficult to obtain otherwise, particularly under simple purchase contracts; and the reciprocal nature of collaboration can bring a degree of incentive capability and stability to the arrangement that would not otherwise be available.

B. Reactive Considerations

The discussion so far has been focused primarily on how external sources of technology can be tapped and what the role of collaborators in this process is.

The strength of the imperative for incumbent firms—i.e., firms currently nicely positioned in the industry—to engage in such activity is more than just a function of the attractiveness of the technological opportunities that lie external to the firm. It may also reflect the fact that failure to shape the new technology may result in the stranding of investments supporting the existing technology. Often this is unavoidable, i.e., a new technology, requiring a new set of inputs and new processing equipment, once commercialized, will destroy the value of investments supporting the existing technology. Incumbents may sometimes be completely helpless before such competitive pressures; however, in some cases, new technologies can be fashioned to deliver superior performance while still placing a demand on the investment put in place to support the old technology. In these cases, affiliation with those developing and shaping new technology has obvious advantages.

C. Implications

The frequency of these various forms of collaboration appear to be increasing. It is not just a feature of new industries like biotechnology, where there are literally thousands of such agreements. It is also a feature of industries like telecommunication equipment and automobiles, and maturing industries like computers and semiconductors (see Mowery, 1988). So great has been the escalation of such activity that one is forced to ask very basic questions about the nature of the firms. Firms are becoming increasingly interconnected through long-term contractual relationships; as the external capabilities of others become increasingly critical to one's own success and as the opportunities for opportunism widen through dependence on fairly loose affiliation of one kind or another, the nature of the firm and the functions of management become transformed.

A number of more specific implications for organizational and business strategy also follow. With respect to incumbents, the analysis suggests that the emergence of technological discontinuities dictates that incumbents must shift gears, latching on to the relevant external sources of know-how, collaborating if necessary to do so. A variety of collaborative modes exist; their respective viability is a function of the degree of protection afforded the technology by intellectual-property protection, and the use of contractual instruments to access the technology in question. Failure to access the new technology may well lead to the demise of the firm. In some cases, this may be the inevitable consequence of a technological discontinuity; more often than not, however, the in-transfer of technology can protect, if not enhance, the competitive standing of incumbents.

Whether collaboration occurs, however, is not just a function of whether willing "buyers" exist. It is also a function of whether the party generating the new technology is willing to collaborate. Generally, there are a variety of factors that encourage them to do so. Besides the obvious infusion of cash that it frequently provides to the source of the innovation, collaboration with incumbents is likely to facilitate commercialization through providing access to complementary capacities in R&D, testing, manufacturing, and marketing.[5] It is to these considerations that we now turn.

V. ACCESSING COMPLEMENTARY ASSETS[6]

In almost all cases,[6] the successful commercialization of new technology requires that the know-how in question be utilized together with the services of other assets. Marketing, competitive manufacturing, and after-sales support are always needed to successfully commercialize a new product or process. These services are often obtained from complementary assets that are often specialized. For example, the commercialization of a new drug is likely to require the dissemination of information over a specialized distribution channel. In some cases, the complementary assets may be the other parts of a system. For instance, hypersonic aircraft may require different landing and servicing facilities.

As a new technology paradigm is developing, usually a number of competing designs are being worked on simultaneously. Before a dominant design emerges, there is little to be gained from firms deploying specialized assets, as scale economies are unavailable and price is not a principal competitive factor. As the leading design or designs begin to be selected by users, however, volumes increase; and opportunities for economies of scale and low cost production will induce firms to begin gearing up for mass production by acquiring specialized tooling and equipment, and possibly specialized distribution as well. Because these investments involve significant irreversibilities, and hence risks, producers must proceed with caution.

The degree of interdependence between the innovation and the complementary assets can, of course, vary tremendously. At one extreme, the complementary assets may be virtually generic, have many potential suppliers including incumbent firms, and be relatively unimportant when compared with the technological breakthrough represented by the innovation. At the other, successful commercialization of the innovation may depend critically on an asset that has only one possible supplier. Such assets might be labelled "bottleneck" assets.

Between these two extremes there is the possibility of "cospecialization"—where the innovation and the complementary assets depend on each other. An example of this would be containerized shipping, which requires specialized trucks and terminals that can work only in conjunction with each other.

A key commercialization decision the owners of the new technology have to make is what to do (build, buy, or rent) with respect to the complementary assets. Although there are a myriad of possible arrangements, two pure types stand out—namely, owning or renting. At one extreme, the innovator could integrate into (i.e., build or acquire) all of the necessary complementary assets. This is likely to be unnecessary as well as prohibitively expensive. It is well to recognize that the variety of assets and competences that need to be accessed is likely to be quite large even for only modestly complex technologies like personal computers.

To produce a personal computer, for instance, a company needs expertise in semiconductor technology, disk-drive technology, networking technology, keyboard technology, and several others. No company has kept pace in all of these areas by itself.

At the other extreme, the innovator could attempt to access these assets through collaborative contractual relationships (e.g., component supply contracts, fabrication contracts, distribution contracts, etc.). In many instances, contracts may suffice, although a contract does expose the innovator to various hazards and dependencies that it may well wish to avoid. An analysis of the properties of the two extreme forms ought to be instructive. A brief synopsis of mixed modes then follows. The perspective adopted is that of the new entrant, rather than that of the incumbent.

A. Contractual Modes

The advantages of collaborative agreements—whereby the innovator contracts with independent suppliers, manufacturers, or distributors—are fairly obvious. The innovator will not have to make the up-front capital expenditures needed to build or buy the assets in question. This reduces risks as well as cash requirements. Also, contractual relationships can bring added credibility to the innovator, especially if the innovator is relatively unknown while the contractual partner is established and viable. Indeed, arms-length contracting that embodies more than a simple buy-sell agreement is becoming so common that various terms (e.g., "strategic alliances," "strategic partnering") have been devised to describe it. Even large companies such as IBM are now engaging in it. For IBM, partners enable the company to "learn things [they] couldn't have learned without many years of trial and error."[7] IBM's arrangement with Microsoft to use the latter's MS-DOS operating system of software on the IBM PC facilitated the timely introduction of IBM's personal computer into the market. Had IBM developed its own operating system, it may have missed the market window.

It is most important to recognize, however, that strategic partnering is exposed to certain hazards, particularly for the innovator and particularly when the innovator is trying to use contracts to access special capabilities. For instance, it may be difficult to induce suppliers to make costly, irreversible commitments that depend for their success on the success of the innovation. To expect suppliers, manufacturers, and distributors to do so is to invite them to take risks along with the innovator. The problem that this poses for the innovator is similar to the problems associated with attracting venture capital. The innovator must persuade its prospective partner that the risk is a good one. The situation is open to opportunistic abuses on both sides. The innovator has incentives to overstate the value of the innovation, while the supplier has incentives to "run with the technology" should the innovation be a success.

In short, the current euphoria over "strategic partnering" may be partially misplaced. The advantages are being stressed (for example, McKenna, 1985) without a balanced presentation of transactional hazards. Briefly, (1) *there is the risk that the partner will not perform according to the innovator's perception* of what the contract requires; (2) *there is the added danger that the partner may imitate the innovator's technology* and attempt to compete with the innovator. Both problems stem from the transactions cost problems discussed earlier. The latter possibility is particularly acute if the provider of the complementary asset is uniquely situated with respect to the specialized assets in question and has the capacity to absorb and imitate the technology.

B. Integration Modes

Integration modes, which by definition involve equity participation, are distinguished from pure contractual modes in that they typically facilitate greater control and greater access to commercial information (Williamson, 1975; Teece, 1976). In the case of a wholly owned asset, this is, of course, rather extensive.

Owning, rather than renting, the requisite specialized assets has clear advantages when the complementary assets are in fixed supply over the relevant time period. It is critical, however, that ownership be obtained before the requirements of the innovation become publicly known; otherwise, the price of the assets in question is likely to be raised. The prospective seller, realizing the value of the asset to the innovator, may well be able to extract a portion, if not all, of the profits that the innovation can generate by charging a price that reflects the value of the asset to the innovator. Such "bottleneck" situations are not uncommon, particularly in distribution.

As a practical matter, however, an innovator may not have the time to acquire or build the complementary assets that ideally it would like to control. This is particularly true when imitation is so easy that timing becomes critical. Additionally, the innovator may simply not have the financial resources to proceed. Accordingly, innovators need to assess complementary, specialized assets as to their importance. If the assets are critical, ownership is warranted although if the firm is cash constrained, a minority position may well represent a sensible tradeoff. If the complementary asset in question is technology or other personnel-related assets, this calculation may need to be revised. This is because ownership of creative enterprises appears to be fraught with hazards as integration tends to destroy incentives and culture (Williamson, 1985).

Needless to say, when imitation is easy, strategic moves to build or buy complementary assets that are specialized must occur with due reference to the moves of competitors. There is no point in moving to build a specialized asset, for instance, if one's imitators can do it faster and cheaper. Figure 3 is a simplified view of how these factors ought to condition the integration decision for a firm that does not already own certain complementary assets needed to bring the new product or process to market successfully.

It is self-evident that if the innovator is already a large enterprise with many of the relevant complementary assets under its control, integration is not likely to be the issue that it might otherwise be because the innovating firm will already control many of the relevant specialized and co-specialized assets. In industries experiencing rapid technological change, however, it is unlikely that a single company has the full range of expertise needed to bring advanced products to market in a timely and cost-effective fashion. Hence, the integration issue is not just a small-firm issue.

C. Mixed Modes

The real world rarely provides extreme or pure cases. Decisions to integrate or license involve tradeoffs, compromises, and mixed approaches. It is not surprising, therefore, that the real world is characterized by mixed modes of organization, involving judicious blends of contracting and integration. Relationships can be engineered around contracts in ways that are functionally akin to integration; internalization can be so decentralized that it is akin to contracts. Still, comparative analysis of the extremes can provide important insights into mixed modes.

Between the extremes of pure contracts and internal organization lie a rich diversity of governance structures that mix elements of both. We view these as an attempt to judiciously combine the flexibility of arms-

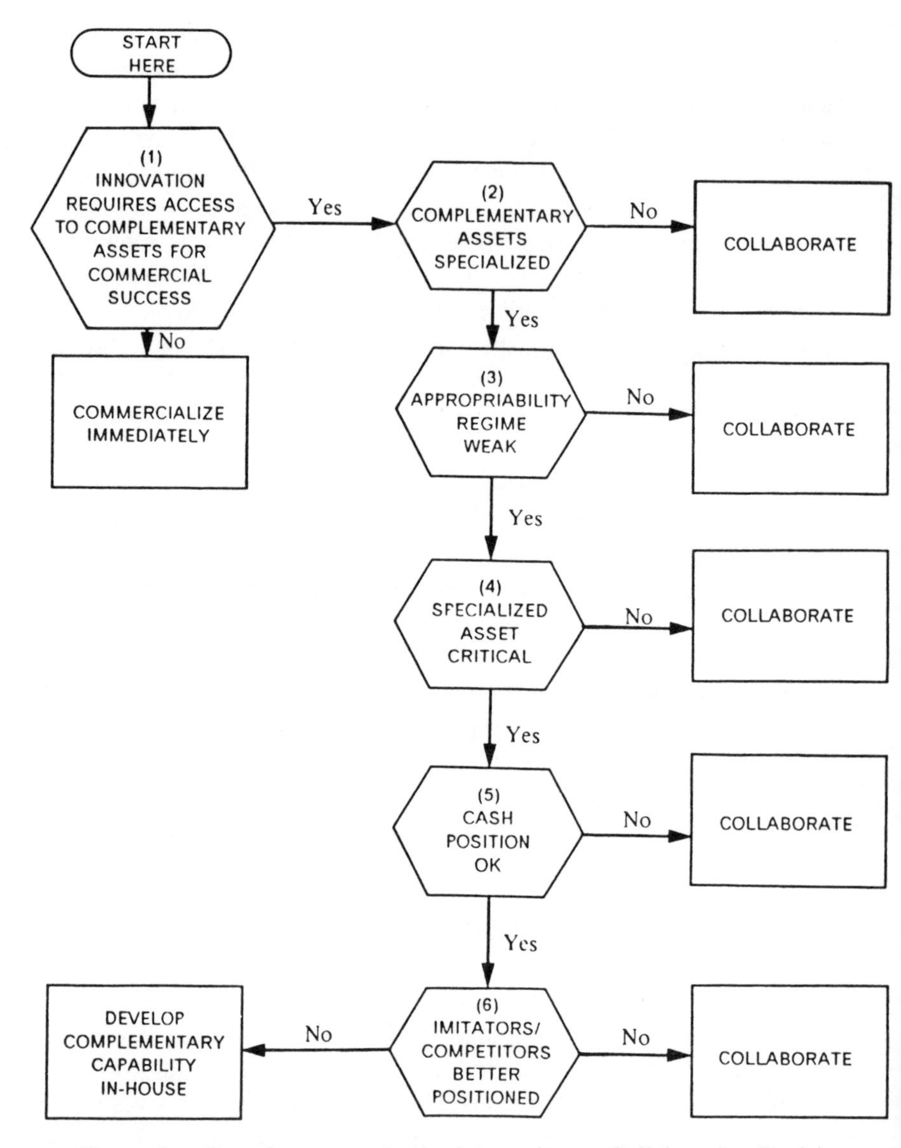

Figure 3. Complementary Assets Integration v. Collaboration Decision

length contracts with the coordination and communication properties of internal organization. There are various mechanisms that can be used to build such intermediate structures. Two of these, equity joint ventures and direct equity participation, are examined below.

Equity Joint Ventures

An equity joint venture, the creation of a new entity jointly owned and operated by the collaborators, is the classic form of organizing collaborative activity. Indeed, most studies of jointly organized activities have focused almost exclusively on this legal form. Equity joint ventures have two governance properties that make them ideal for coordinating complex transactions involving specialized assets. First, they create an administrative hierarchy (quite similar to internal organization) for setting general operational and strategic policies as well as for settling disputes. This hierarchical structure dispenses with the need for collaborators to attempt the often impossible task of specifying a complete set of contractual provisions for conducting the collaboration. Instead, the collaborators need only agree on a broad set of governing rules that provide a framework for deciding on more specific conditions as more information becomes available. In addition, the governing body of the venture, usually composed of representatives of both companies, provides a channel for communicating pertinent information and for coordinating the collaborative roles of each partner.

The second advantage of equity joint ventures is that both parties hold a direct stake (through their equity position) in the success of the project. This feature aligns incentives and can lower the risk that one party will become opportunistic. Partners pay some share of the costs of any actions they take that hurt the viability of the venture. In addition, the formal ownership structure provides each party legal rights with respect to the technology of other strategic assets contributed to or developed by the venture. Parties can agree at the outset about the division of assets if the venture is terminated.

Joint ventures also entail certain costs that must be recognized. Generally, they take longer to negotiate and organize than other, less hierarchical forms of governance. Given these costs, they are usually appropriate only for longer-term projects that involve heavy capital or technological commitment by both parties.

Direct Equity Positions

An alternative to establishing a jointly owned company is for one partner to take a direct equity stake in the other. This is often used where a significant size differential exists between the collaborators, and it would be impossible for the smaller party to contribute enough equity into a jointly owned company. The direct-equity approach is similar to that of equity joint ventures, although generally providing for less joint control. First, the equity stake again helps to align incentives. It safeguards the smaller partner (or investee) by creating direct costs for the investor to act opportunistically. If the inventor takes any action that hurts the investee, it will bear some portion of the resulting costs through its equity stake. Usually, at the time the equity investment is made, the parties concurrently agree on a set of longer-run strategic and operational goals of the relationship. The contribution of equity helps to ensure that the investor will have an interest in ensuring that these strategic and operational goals are pursued in good faith.

Second, the direct-equity stake can provide some scope for hierarchical governance (as opposed to strictly contractual governance) if it allows the investor a seat on the other company's Board of Directors. The goal is generally not to achieve voting power. Instead, it is to gain a direct communication channel to the highest governing level of the other partner. This ensures that the top management of the partner stays interested in the business relationship. It can also help to ensure that critical problems and issues will be brought directly to the top management, rather than having to percolate up from the line managers in charge of the collaborative effort. The board position also helps information to circulate the other way, from the investee to the investor. Often, the corporate investor appoints one of its high-level executives to fill the board seat and thus provides the investee with a direct channel back to the corporate partner. Like the equity joint venture, direct equity relationships have advantages over non-equity, contractual forms of collaboration when the activities in question involve transaction-specific assets and uncertainty.

Sometimes mixed modes represent transitional phases. For instance, because of the convergence of computer and telecommunication technology, firms in each industry are discovering that they often lack the requisite technical capabilities in the other. Because the technological interdependence of the two requires collaboration among those that design different parts of the system, intense cross-boundary coordination and information flows are required. When separate enterprises are involved, agreement must be reached on complex protocol issues among parties that see their interests differently. Contractual difficulties can be anticipated because the selection of common technical protocols among the parties will often be followed by transaction-specific investments in hardware and software.

The use of contractual, equity, and internal forms of organization in the telecommunications-equipment industry are discussed in the following section.

VI. THE GOVERNANCE OF COLLABORATIVE RELATIONSHIPS IN THE WORLD'S TELECOMMUNICATIONS EQUIPMENT INDUSTRY

In this section, we apply the above framework to examine collaborative arrangements in the telecommunications equipment industry. We have chosen to study collaboration in telecommunications because fundamental changes in equipment technology and the structure of demand have altered the bundle of strategic assets necessary to compete. This situation has created a *market for strategic assets* among firms with distinctive technological and commercial competencies.

Consistent with the framework presented above, we can divide strategic assets in the telecommunications equipment industry into two broad categories: (1) upstream technical capabilities and know-how needed to develop complementary or critical input technologies; (2) downstream marketing and distribution needed to penetrate particular product, customer, or geographic markets. These strategic assets are traded through a rich variety of governance structures including arm's length contracts, equity joint ventures, and partial-equity linkages.

We focus our analysis on arrangements where firms collaborate in the development or exchange of technical know-how. These include cases where technology transfer or joint technological development form the sole functional basis for collaboration *and* cases where these technological functions are coupled to downstream functions (such as manufacturing and marketing) in a single collaborative relationship. There are several examples which also illustrate the varied motives behind such collaborations.

First, one motive for collaboration is legal. Firms with overlapping technologies may strike a licensing deal to avoid or settle patent litigation. These agreements, which are quite common in semiconductors (Malerba, 1985), generally involve cross-licensing and consist of little more than a grant of permission to use some proprietary technology.

A second motive, which goes beyond sheer legal considerations, is to tap the specialized technical expertise of another firm. In recent years, the functions of communications equipment have expanded rapidly. Traditionally, telecommunications equipment was designed for voice communications. Increasingly, telecommunications also encompasses the transmission of data, text, image, and video. This functional expansion has made it necessary for communications equipment to incorporate a far broader range of component and sub-system level technologies. As the range of technologies underlying telecommunications systems has expanded, it has become increasingly inefficient for telecommunications equipment firms to track all of the relevant technological frontiers through in-house R&D. Access to state-of-the-art technologies is often better served by collaborating with specialized producers of components (e.g., semiconductors) and sub-systems. As part of the functional expansion of telecommunications, equipment suppliers must ensure that these products are compatible with the various types of new terminal equipment (computers, facsimile devices, voice messaging devices, etc.) which have become part of telecommunications networks. To ensure equipment compatibility, firms license communications protocols and jointly develop complementary systems technologies. In all of the above cases, collaboration can stop at the technology stage or encompass downstream functions such as manufacturing and marketing.

We are interested in the organization and governance of these technology-based transactions. In particular, we are interested in the choice between purely contractual arrangements (which include both short- and long-term contracts) and what we call quasi-internal arrangements (equity joint ventures, direct investments, and organized consortium). The governance properties of equity joint ventures and direct equity relationships were discussed in Section V.A. Our general hypothesis is that these quasi-internal structures can more efficiently govern technical collaborations involving durable, transaction-specific investments and uncertainty. A test of this hypothesis is presented in the section below.

A. Source of Data

The source of data for our analysis is a database constructed by researchers at Futoro Organizzazione Risorse (F.O.R.) in Rome through an extensive review of the trade press.[8] A total of 974 collaborative arrangements were recorded over the period 1982–1985, 117 of which were in telecommunications equipment. Among the information recorded for each arrangement was the legal form, function (or motive), respective nationalities of the collaborators.

Construction of the Dependent Variable

To construct the categorical dependent variable, we used information on the legal form of each arrangement. These were classified as follows:

1. *Nonequity Agreements:* Agreements which do not include equity ties or the creation of new, jointly owned companies. Examples included contracts involving various activities and licensing agreements.
2. *Equity Agreements:* Agreements where one party acquires a minority equity interest in another for some industrial purpose (i.e., excluding purely financial investments).
3. *Joint Ventures:* Agreements in which a new legal entity is created and jointly owned by the partners.
4. *Consortia:* Joint ventures involving more than one partner.

These categories could be translated directly into our binary dependent variable classification scheme. The category ''non-equity agreements'' formed our ''purely contractual governance'' category while equity-agreements, joint ventures, and consortia were classified into our ''quasi-internal'' governance category.

Independent Variable

Constructing independent variables to reflect transaction specificity and uncertainty proved far more difficult. We had data neither on the levels of transaction-specific investments nor on the levels of uncertainty involved in the collaborative projects. We were therefore forced to use the functional purposes of the arrangements as a categorical indicator of both uncertainty and transaction-specificity. In the database, each arrangement was classified into one of the following nine functional purpose categories:

1. *Technology-transfer:* A unilateral assignment of licenses, which may also be accompanied by technical assistance.
2. *R&D Integration:* Cross-licensing, the provision of R&D services, and joint R&D for the development of new products.
3. *Supply Arrangements:* Agreements for the provision of goods, either short-term or long-term.

4. *Production Integration:* Joint production of intermediate or finished goods.
5. *Distribution/Marketing Integration:* Sales and marketing by one party for another or jointly by both parties.
6. *Integration of R&D and Production:* Combination of 2 and 4 above.
7. *Integration of R&D and Distribution/Marketing:* Combination of 2 and 5 above.
8. *Integration of Production and Distribution/Marketing:* Combination of 4 and 5 above.
9. *Integration of R&D, Production, and Distribution/Marketing:* Combination of 2, 4, and 5 above.

Our interest lies with those collaborations involving technology transfer or R&D integration as at least one functional motive. As discussed earlier, R&D collaboration and technology transfer may involve transactional difficulties. Such difficulties are likely to be compounded when collaboration couples technological activities with commercialization functions (production, distribution/marketing). Attempts to jointly create *and* utilize new technology require more communication and coordination between collaborators than when collaboration is limited to R&D. In addition to agreeing on technological goals and tactics, collaborators will also have to coordinate investments in production facilities and decisions on marketing/distribution. Governing the *link* between R&D and commercialization also imposes additional organizational burdens on collaboration. The hazards associated with transaction-specific capital are likely to be greater when partners must make investments in costly and durable downstream assets needed to utilize the know-how in question. For the above reasons, we expect collaborations that involve *both* technological activities and downstream functions to have a greater likelihood of being organized through quasi-internal governance structures than collaborations that only involve technological activities (i.e., technology transfer or R&D integration).

The raw data on function and form are summarized in Table 1. These data reflect the rich diversity of collaborative functions and forms through which collaboration takes place.

Methodology

To examine the hypothesis that technology collaborations that also involved downstream functions would be more likely to be organized

Table 1. Frequency of Form and Function of Telecommunications Cooperative Agreements

Function	*Nonequity*	*Equity*	*Joint Venture*	*Consortium*	*Total*
1. Technology Transfer	5	4	0	0	9
2. R&D Integration	23	2	1	1	27
3. Supply Arrangements	9	0	0	0	9
4. Production Integration	2	0	6	7	15
5. Distribution/Marketing Integration	26	1	8	0	35
6. R&D, Production Integration	2	0	5	0	7
7. R&D, Distribution/Marketing Integration	5	1	2	1	9
8. Production, Distribution/Marketing Integration	2	0	1	0	3
9. R&D, Production Distribution/ Marketing Integration	1	0	2	0	3
Total	75	8	25	9	117

under quasi-internal arrangements than when they did not include downstream activities, we conducted chi-square tests for homogeneity.[9] The null hypothesis in these tests is that a particular functional type of collaboration (e.g., R&D alone) will be organized under a particular governance form (e.g., quasi-internal) with roughly the same relative frequency with which another function is organized under the same form. For example, if in a sample of collaborative arrangements, quasi-internal governance forms represent 20% of arrangements, there would be a 20% probability that any given functional type of arrangement within that sample would have a quasi-internal governance form. Systematic deviations from the expected probability distribution should be explained by theory. The null hypothesis represents a view, contrary to our own, that there is no systematic relationship between the governance category and the functional category of collaborations. Before proceeding, the reader should be aware of the following caveats.

First, while we have been fortunate enough to have access to an extensive database of collaborative arrangements, it undoubtedly contains some biases based on the primary sources (i.e., the business press) from which it was gathered. The business press probably has a greater tendency to report collaborations struck between the major players in the marketplace. In addition, the press can cover only those arrangements that are not purposely kept secret by the collaborators. Generalizations based on our results should be limited to the relevant sampling frame (i.e., cases where an arrangement was publicly announced and where it involved a firm or project that was considered "newsworthy").

Second, we must stress another, more serious form of sample-selection bias inherent in this type of analysis. Our data includes only those cases where some form of collaboration actually took place: we do not (and cannot) observe those cases where it did not. In statistical terms, we have censored cases. The sample distribution is biased because it does not include those unobservable cases where firms did not choose internal organization over collaboration. This is particularly relevant to our analysis if a firm chose to organize a particular project internally due to the high transaction costs of collaboration. Unfortunately, the structure of our data does not permit us to apply any of the available statistical techniques to correct the problem.[10]

Finally, we must note that the information contained in each arrangement permits us to make judgments neither about the uncertainty nor about the transaction-specific investments involved in each. In lieu of such detailed information, we have relied on our knowledge about the nature of technology transfer, R&D, production, and distribution for telecommunications equipment. This approach obviously limits the degree to which we can make predictive statements about the relationship between form and function.

Despite these limitations, we think that our analysis is an important step forward in the study of collaborative organizations. It is one of the few attempts to analyze statistically the relationship between the form and function of collaborative arrangements within a transaction-costs framework (see also Pisano, 1988). The results are intended to stimulate further analysis along these lines.

Analysis and Results

Because our interest centers on technology-based collaborations, the relevant sample consists of those cases where R&D or technology-transfer represents *at least one function* (and perhaps the only function) in a collaborative arrangement. There are 55 arrangements that fit this criteria. Of these 55, 36 were cases where R&D integration or technology-transfer formed the sole functional basis of collaboration. For convenience, we will refer to this type of arrangement as "technology-only." The remaining 19 cases included collaborations that combined a technological function with either manufacturing, marketing, or both. We

Table 2

Function	*Technology-only*	*Technology-plus*	*Total*
Form:			
contract	28 (23.56)	8 (12.44)	36
quasi-internal	8 (12.44)	11 (6.56)	19
total	36	19	55
chi-square = 7.01*			

*results significant beyond .01 level

will refer to this category as "technology-plus." The technology-plus category was an aggregation of three types of functional cases found in the database: R&D plus manufacturing, R&D plus distribution, and R&D plus manufacturing and distribution.

Table 2 below shows the results of a contingency table analysis that compares the distribution of governance structures for technology-only and technology-plus arrangements. The expected frequencies for each cell are shown in parentheses.

The results allow us to reject the null hypothesi (at $p < .01$) that there is no systematic relationship between the functional category (technology-only vs. technology-plus) and the chosen governance structure (contract vs. quasi-internal). Comparing the actual frequencies with the expected frequencies in Table 2 also indicates that an independence model underpredicts the actual frequency with which technology-plus arrangements are organized under quasi-internal governance structures (11 actual cases versus 6.56 predicted cases). These results suggest that the actual frequency with which technology-plus arrangements are organized under quasi-internal governance structures can not be explained by purely random forces. This suggests that substantive differences between the two functional categories explain the relative difference in the frequency with which one type of governance structure is chosen over another.

One potential problem in our analysis is that the technology-plus category was formed by aggregating three different types of arrangements. This assumes that the three underlying functional types constitute a homogeneous population with respect to governance choices. We tested whether there was a basis for this assumption with chi-square tests for homogeneity. The results of these tests are provided below.

These results do not allow us to reject the null hypothesis that these

Table 3

Function	*R&D/MFCT and R&D/MARKETING*	*R&D/MFCT/MKT*	*Total*
Form:			
contract	7	1	8
quasi-internal	9	2	11
Total	16	3	19
chi-square = .11			

Function:	*R&D/MFCT*	*R&D/MARKETING*	*Total*
Form:			
contract	2	5	7
quasi-internal	5	4	9
Total	7	9	16
chi-square = 1.17			

functional categories are relatively homogeneous with respect to governance choices. The basis for aggregation appears to be sound.

VIII. CONCLUSIONS

In this paper a theory of the organization of R&D has been developed which we believe has normative implications for technology strategy. Technology strategy we argue involves not only an understanding of the commercial significance of technological development but also an understanding of how best to organize to take advantage of technological and commercial opportunities. Our data from the telecommunications equipment industry indicated that the exchange, development, or commercial exploitation of technology is a frequent motive for collaboration. However, the case of telecommunications equipment is not unique. Technological motives are a frequent motive for collaboration in industries such as biotechnology (Pisano, Shan, and Teece, 1988), semiconductors, robotics, and computers (see Mowery, 1988).

Our data analysis, while limited in many respects, suggested that the governance of collaborative arrangements is related to the function of collaboration. When collaboration is designed to couple technology and some downstream activity, there seems to be a tendency to go beyond pure contracts as a governing mechanism.

There are two major limits to our analysis which we hope that future research will address. First, our data did not allow us to examine internal vs. external choices. Secondly, data limitations prevented us from testing the normative aspects of our framework. Future empirical research should seek to link performance outcomes to governance choices (including internal organization) under a variety of conditions. These results will be of most interest to corporate managers whose responsibilities will increasingly involve choices about the appropriate organization to both exploit and acquire new technologies. Policymakers can also benefit from an understanding of how firms organize for innovation, and the properties of alternate structures.

ACKNOWLEDGMENT

We wish to thank Mel Horwitch and Mike Russo for helpful discussions on the subject matter of this paper. Research support from the NSF, grant # SRS-8410556, is gratefully appreciated.

NOTES

1. Needless to say, the corollary for the new technology-based business firm is whether it should license its technology to an established firm or attempt to commercialize it internally.
2. Horwitch (1988) refers to these as "post-modern" firms.
3. For a similar view, see Horwitch (1988, chapter 4).
4. This poses the issue in static terms. It may well happen that external sourcing strategies adopted in the past may deny the firm the ability to develop, at competitive cost, technologies relevant to today's market necessities.
5. If the source of the technology already owns the relevant assets, or has the capacity to build them, its motivation to collaborate will obviously be attenuated.
6. This section is based in part on Teece (1986, forthcoming).
7. Comments attributed to Peter Olson III, IBM's director of business development, in "The Strategy Behind IBM's Strategic Alliances," *Electronic Business*, October 1, 1985, p. 126.
8. We are deeply indebted to Enrico Ricotta for providing us with access to this data.
9. See Feinberg (1977) for a rigorous treatment of contingency table analysis.
10. For a discussion of the censored and truncated distributions and how to deal with them, see Maddala (1983), Chapter 6.

REFERENCES

Arrow, K., 1962. "Economic welfare and the allocation of resources for invention," in R. Nelson (ed.), *The Rate and Direction of Inventive Activity*. Princeton: Princeton University Press.

Feinberg, S., 1977. *The Analysis of Cross-Classified Categorical Data*. Cambridge: MIT Press.

Foster, R., 1986. "Timing Technological Transitions," in Mel Horwitch (Ed.), *Technology in the Modern Corporation*. NY: Pergamon Press.

Horwitch, M., 1988. *Post-Modern Management*. New York: Free Press.

Levin, R., A. Klevorick, R. Nelson, and S. Winter, 1984. "Survey Research on R&D, Appropriability, and Technological Opportunity." Unpublished manuscript, New Haven, CT: Yale University.

Maddala, G., 1983. *Limited-Dependent and Qualitative Variables in Econometrics*. Cambridge: Cambridge University Press.

Malerba, F., 1985. *The Semiconductor Business*. London: Frances Pinter.

McKenna, R., 1985. "Market Positioning in High Technology," *California Management Review*, 27, 3, (Spring).

Mowery, D. (ed.), 1988. *International Collaborative Ventures in U.S. Manufacturing*. Cambridge, MA: Ballinger Publishing Company.

Nelson, R., & S. Winter, 1982. *An Evolutionary Theory of Economic Change*. Cambridge, MA: Harvard University Press.

Pisano, G., 1988. *Innovation Through Markets, Hierarchies, and Joint Ventures: Technology Strategy and Collaborative Arrangements in the Biotechnology Industry*. Unpublished Ph.D. thesis, University of California, Berkeley, School of Business Administration.

Pisano, G., W. Shan, and D. Teece, 1988. "Joint Ventures and Collaboration in Biotechnology," in D. Mowery (ed.), *International Collaborative Ventures in U.S. Manufacturing*. Cambridge, MA: Ballinger Publishing Company.

Rothwell, R. & C. Freeman et al., 1974. "SAPPHO Updated-Project SAPPHO Phase II," *Research Policy* 3, pp. 258–291.

Teece, David J., 1977. "Technology Transfer by Multinational Firms: The Resource Cost of International Technology Transfer," *Economic Journal*, (June).

Teece, David J., 1981. "The Market For Knowhow and the Efficient International Transfer of Technology," *Annals of the Academy of Political and Social Science*. (November).

Teece, David J., 1986. "Profiting from Technological Innovation," *Research Policy* 15, No. 6, (December) pp. 285–305.

Teece, David J., forthcoming. "Market Entry Strategies for Innovators," *Journal of Strategic Management*.

Von Hippel, E., 1982. "Appropriability of Innovation Benefit as a Predictor of the Source of Innovation," *Research Policy*, 11:2, (April) pp. 95–115.

Williamson, Oliver E., 1975. *Markets and Hierarchies*. New York: Free Press.

Williamson, Oliver E., 1985. *The Economic Institutions of Capitalism*. New York: Free Press.

Winter, S., 1984. "Schumpeterian Competition in Alternative Technological Regimes," *Journal of Economic Behavior and Organization* 5, No.3–4, (September–December) pp. 287–320.

ASSESSING THE SCOPE OF INNOVATIONS:

A DILEMMA FOR TOP MANAGEMENT

C. K. Prahalad, Yves Doz, and R. Angelmar

The scope or the domain of a business is at the very heart of strategic analysis. Questions such as "what business are we in" are intended to force managers and analysts to circumscribe the boundaries or the domain of a business. It is no surprise, therefore, that defining the scope of a business has been an integral part of the strategy formulation process.[1] While domain definition has always been a critical component of strategy formulation, few studies exist in the literature on the criteria to be used for domain definition or the internal organizational processes that influence it. Abell initiated this line of enquiry.[2] His study provided a framework to define scope by identifying a few criteria that one could profita-

Research on Technological Innovation, Management and Policy
Volume 4, pages 257–281.

ISBN: 0-89232-798-7

bly apply—customer groups, customer functions and technologies. Porter linked competitive scope definition to the study of the value chain. He also elaborated on segmentation—an approach to domain definition.[3] Normann[4] and Miles and Snow[5] provided a framework that linked "domain definition" with its organizational implications. Abell went one step further and recognized that the domain definitions can be *fragile* and that managers must become sensitive to the changing concepts of the business brought about by new technology and/or evolving functionality. Domain definition analysis, he argued, must be dynamic and on-going.[6] While Abell recognized the fragility of domains, technological advances during the last decade are introducing a qualitatively different problem—a scope of a business which is underdetermined, ambiguous, and often times unpredictable. Consider for example:

1. The evolution of information technology—computers and communications—for instance, allowed Merrill Lynch to create an integrated service business called the "Cash Management Account." It integrated the services and businesses which had previously been quite distinct; the business of a bank and the brokerage house. In this case, it was a case of combining the domains of two distinct businesses brought about by two technologies which were converging.
2. The emergence of digital technology has resulted in a blurring of lines between business segments. For example, is the personal computer a consumer or a professional business? While marketing considerations such as choice of channels, pricing, sales and service support may be different between consumer and professional segments, the basic functionality, the manufacturing and the product development process may be the same. Various firms entering the PC business in 1982 approached the business differently, reflecting this dilemma.[7] Should manufacturing and technology be considered to be criteria for establishing a business domain? This dilemma forces us to reconsider our concept of a domain. Traditionally, domain of a business was defined using customer related criteria such as customer groups, and functionality. We have not considered, explicitly, business domains that are determined by the very nature of the technologies. For example, should *core technologies* be seen as defining the appropriate business domain of the company? These core technologies, either independently, or in combination with other technologies may provide the opportunity for the company to participate in a very large number of product market domains. While the business may be quite distinct at the product market level, they may all represent an

extension of a core set of skills. Canon, for example, participates in businesses as diverse as cameras, calculators, copiers, facsimile, personal computers, and printers. They are quite distinct businesses. However, underlying these businesses are a set of core competencies in imaging, sensors, displays, miniaturization, microprocessors, optics, power sources, and materials. Should business domains be defined using core technological competencies as well? Itami, for example, considers "technological fit" as distinct from "customer fit" in discussing strategy choices.[8] More on this later.

3. Business domains are radically altered by technologies that are quite far removed from the base business. For example, the film distribution business, worldwide, has been dramatically altered by the emergence of the video recorder (VCR) and the attendant surge in prerecorded software. Instead of depending solely on distribution through movie houses, feature films can now be sold or rented as video cassettes. The video rental business has emerged as a substitute for the traditional cinema theatre business. In the case of the toy industry, the emergence of the cheap microprocessor, has created a whole new segment—video games. While the traditional toy industry is still active and growing, grafting of a new technology far removed from traditional toy industry "technologies" has revitalized it. The role of pacing technologies in changing staid businesses as well as creating new ones is shown schematically in Exhibit 1. What process of analysis would have allowed managers in the film distribution or the toy industry to anticipate these changes?

These examples would suggest that technological changes are the primary drivers creating ambiguity to the definition of domain of a business. Surprisingly, the literature on innovation is silent on the issue. There has been very little attention paid to the process by which the "scope" of an innovation is determined in a firm. The literature on the process of innovation and new product development in a firm appears to take scope of the innovation (and therefore the domain of the business) for granted. The research attention is on a relatively focused and self contained view of innovation such as product development. This view assumes that the problems of domain definition of the business and scope of the innovation have been resolved, or at least reduced to its market segmentation level.[9]

Definition of a domain is qualitatively a different problem in start-up ventures compared to large, multibusiness firms involved in similar technologies. Often, while the start-up venture may be involved in technologies that can potentially influence several existing industries or spawn a

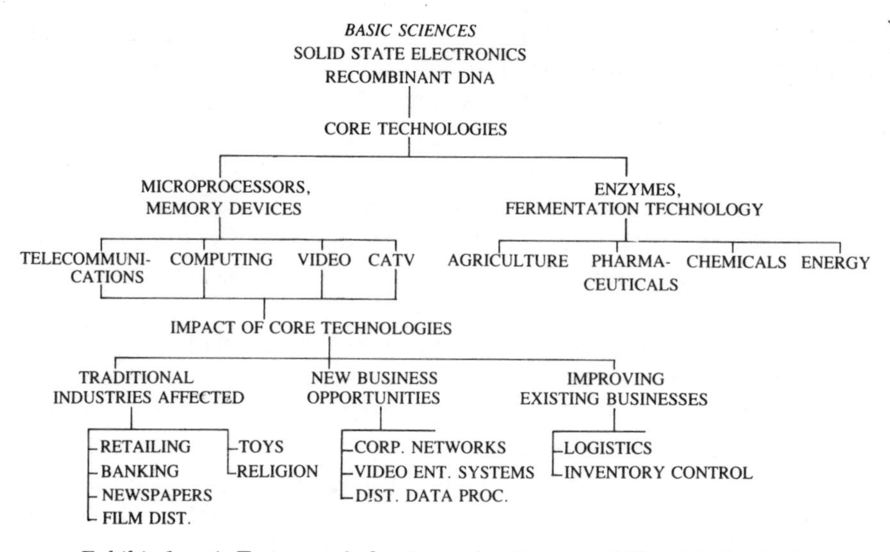

Exhibit 1. A Framework for Assessing Impact of New Technologies

variety of new industries, the resource limitations and the unique and often idiosyncratic perspective of the founding group can provide a focus, even though that definition of domain may be quite arbitrary. On the other hand, in a large multibusiness firm, the definition of domain can become a bone of contention among managers charged with responsibility for different businesses. The literature on corporate venturing somehow sidesteps the issue. The literature focuses more on the process and less on the substance of the underlying technological and market forces, other than using relatively abstract concepts such as "familiarity" as guidance for the venture process. According to this literature, senior management treats the issues of scope of the business as an after thought. The new venture, it is argued, can be regrouped with an existing business or treated as a new business depending on the view of senior management at that point in time. This is the task that is often described in the literature as the creation of the "strategic context" for the innovation. Managers are described as treating the innovation process as an "autonomous process" even when it takes place in large firms. The creation of "strategic context" for a new business involves the gradual integration of the "autonomous" strategic thrust at the level of individual businesses with the corporate strategy. In particular this process leads to a definition of scope for the new business. Yet, this process, as described by Burgelman[10], leads to changes in the scope of a business being achieved by autonomous strategic behavior on the part of lower level managers whose choices are later recast into the current top management vision of business strategy. While this process may be effective in developing new businesses in areas not directly related to current operations, the issues we deal with in this paper are different. We are not considering how top management can rationalize ex-post "autonomous" strategic development at lower levels, but how innovation in core technologies that affect multiple businesses (either because they draw on their knowhow or find application in several existing business domains) can be steered to best serve corporate interest rather than be left—or fall prey—to the parochial interests of individual business units.

In this paper we focus on the problem of defining the scope of innovations in large, multibusiness firms. The task of defining scope is not a one time effort. As the technology evolves, the scope of the innovation has to be continuously monitored and readjusted. We will therefore take a dynamic view of domain definition. We will first illustrate the problem of domain definition via three "mini-cases." These summarized cases are not used to document descriptively the actual choices made by one or another company, but to demonstrate the significance and difficulty of domain definition issues. Subsequently, we discuss the conditions that make domain definition particularly difficult, followed by an analysis of the key management processes directly impacted by the difficulty. The paper concludes by outlining the organizational capabilities necessary for effective scoping of innovations.

I. ILLUSTRATION OF THE PROBLEM

In this section we will illustrate the nature of the problem and the dilemmas that top managers face in large, multi business firms. Scoping an innovation becomes a problem whenever new knowledge and/or competency becomes available to the large firm. Consider the following examples:

Case 1: The Emergence of Optical Media

The use of optical media to store digital information is a relatively new technology. It was popularized by the compact audio disc, jointly developed by Philips and Sony. The compact audio disc has had spectacular success as a substitute for the audio cassette and the vinyl discs as the quality of sound reproduction is considerably better in compact discs. The underlying technology is based on digitalizing sound (converting it into binary information) and burning holes into a disc to contain that information. The compact disc technology, while developed initially to be a commercial success as an audio product, is essentially a method for storage of data. The technology provides an opportunity to store and selectively retrieve very large amounts of data. As a method of storing data, it may develop into many other applications, both consumer and professional, as shown in Exhibit 2. As can be seen from the exhibit, the business opportunity can be identified as (a) extensions of the audio application—car entertainment systems, a portable compact audio disc "walkman" etc., (b) exploring the possibility of video applications or storing pictures in the same format. Such an application, popularly called "laser vision," is already available, (c) creating a possibility of combining audio-visual applications for the consumer market, (d) exploring the development of new applications like satellite based navigation of automobiles, (e) explore the possibilities in the application of this technology to create computer peripherals (disc storage) as competition to magnetic storage technology, (f) use the technology to revolutionize several staid industries like book publishing (one disc could contain all the information in the Encyclopedia Britannica), and (g) create relatively indestructible archival systems, necessary in such applications as bills of materials for weapons systems. The opportunities seem limitless. Optical media, as a technology opens up a whole host of opportunities for companies such as Philips, Sony, Matsushita, Toshiba and others who are already strong in consumer businesses (audio and video products) and have respectable market shares in professional businesses such as office automation. These firms can treat the compact disc as a technology to be nurtured and exploited in an "audio" product division or group. It was this business that gave the initial commercial momentum to the technology. However, should the scope of this innovation be restricted to audio products, especially in firms which participate in several markets where this technology could have major impact and can help revitalize them? Should top managers allocate resources to optical media in the context of the audio business or to the development of optical media as a distinct entity bearing in mind the potential benefits of this technology to several existing businesses in their portfolio as well as the possibility of developing new businesses?

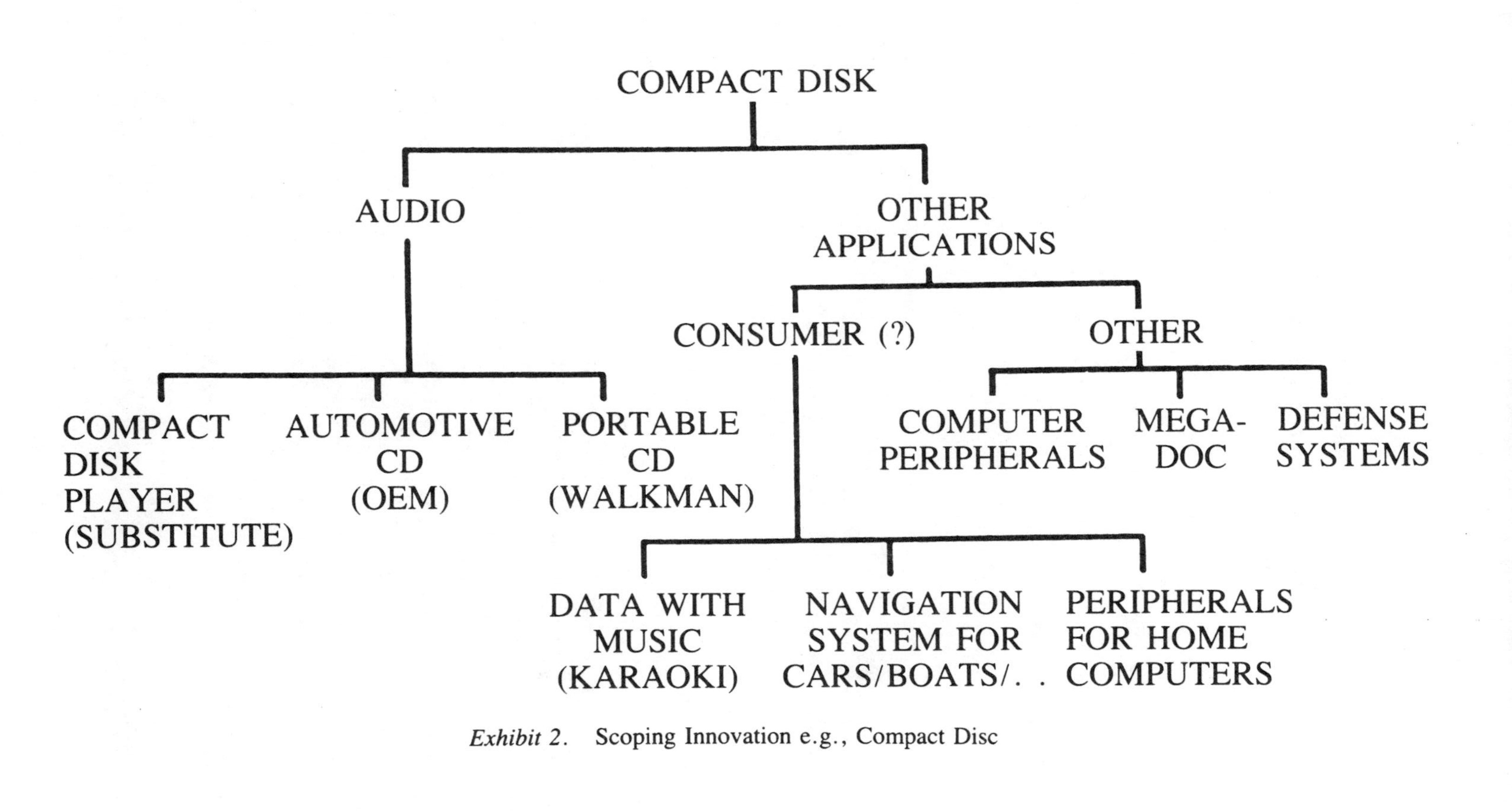

Exhibit 2. Scoping Innovation e.g., Compact Disc

Case 2: The Video Camera Dilemma

The most exciting consumer electronics product to hit the market during the period 1979–82 was the video recorder. Popularly known as the VCR or VTR, it added a whole new dimension to the television screen from being a passive instrument to one where the consumer has some control on what he/she wants to watch, through "time shifting" (recording regular television programs and watching them later at one's convenience) or using "pre-recorded software." The next stage in personalizing the television is to use the video camera which allows one to make one's own movies and watch them using a television set in conjunction with a VCR. The videocamera business has not become as much of a success as the VCR yet due to two reasons: (a) the weight of the camera has been such that it might be really considered "transportable" rather than "portable." There has been a significant effort to reduce the weight without sacrificing quality and/or features, and (b) firms had to decide which tape format they should use. While in the VCR business, the VHS format developed by JVC and Matsushita has become the clear winner over the Betamax format of Sony and the V-2000 of Philips, in the videocamera business, there was an alternate format that was possible called the "8 mm format," a new industry standard. Some firms, such as Sony, have been pushing the "8 mm format." The 8 mm format provided an opportunity for weight and size reduction of the video camera, thus solving one of the problems identified as an impediment to rapid diffusion of that business. The dilemma for top management was the following: The growth of videocamera business was expected to be a function of the installed base of VCRs. The format issue (VHS or Betamax or 8 mm) was a critical issue. Further the number of VCRs was a function of the installed base of television sets in a market. This provided an interesting dilemma to top managers. Should we focus on the videocamera business or on other possible additions to the installed base of television sets by emphasizing the home information systems built around the TV such as videotext, and home security systems rather than emphasizing the home entertainment system which videocamera represented. The market evolution and emphasis could be totally different depending on which view prevailed. On the other hand, if the video camera business ought to be pursued, the technical folks in these firms came with a whole new set of tradeoffs. The emerging technology of electronic imaging, represented by charge coupled devices (CCD), provided a new opportunity to both reduce weight, cost, and improve picture quality simul-

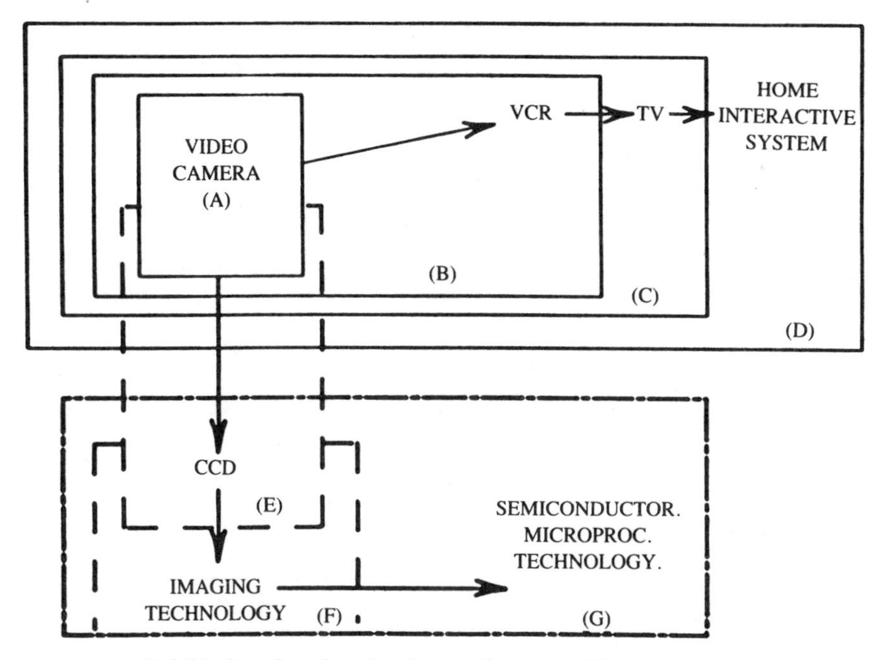

Exhibit 3. Scoping the Innovation e.g., Video Camera

taneously. What is more, it also allowed the firms to sharpen their capability in semiconductor technologies, which are as fundamental as the optical media described in Case 1. The marketing and the technological debate in firms can be visually represented as shown in Exhibit 3. Again, top managers have to contain the scope of the innovation. However, the logic of the arguments from both related marketing groups (VCR, TV, Videotext. . . .) and technical groups (semiconductors) could not be ignored.

Case 3: Packet Switching

Packet switching was a new concept in telecommunications around 1978–1980. It evolved around the early 1970s, as a project for the DOD. Key members of the team who were involved in the project started a private firm, called Telenet, to offer the service to the private sector. As a stand alone business, the scope of packet switching pioneers, such as Telenet, was restricted to providing the "communications pipeline." Customers decided the nature of applications that they used the "pipeline" for. Intrigued by the success of packet switching technology,

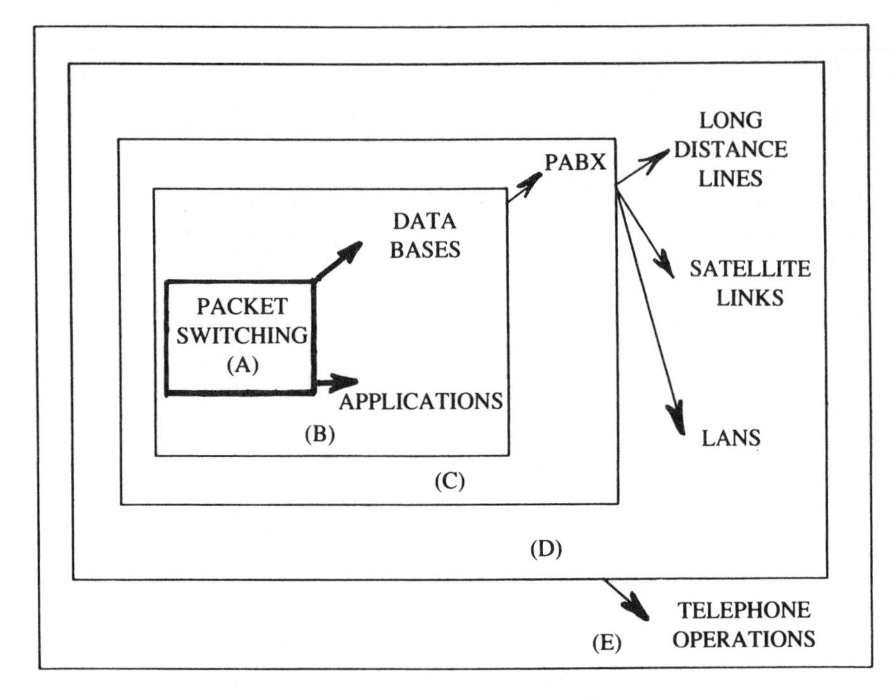

Exhibit 4. An Example of the Evolution of Business Domain.

large telecommunications companies like AT&T and GTE either initiated their own efforts in packet switching or acquired existing small firms like Telenet. The scope of the packet switching business, in the context of the large communications firm, started to evolve. Initially, application packages like electronic mail and data bases like financial and stock market data were added. Subsequently, packet switching was seen as an integral part of the digital PABX business and office automation. Finally it became a part of the total telecommunications capability of the whole company—including telephone operations, and indistinguishable from other businesses in the portfolio. Technologically, this evolution can be justified. In a business sense, the underlying economics of each of the "component parts" of the total telecommunications systems portfolio of the large firm was so distinct that bundling all of them as a part of a total system provided no clarity to the competitiveness of the parts. The evolution of the domain of the packet switching business in one large firm is shown in Exhibit 4.

We could multiply examples such as these. These changes in the domain definition took place, in the cases we studied, not over a very long period of time, through a process of trial and error, or imitation of competitors, or at the request of key customers. These changes in the domain of the innovation, shown in Exhibits (2), (3), and (4) took place within a period of three to four years. What are the characteristics of these innovations and the characteristics of large organizations that lead to this constant redefinition of the domain of a business?

II. AMBIGUOUS BUSINESS DOMAIN DEFINITIONS

The three examples we have cited above are representative of generic situations where the definition of a business domain is fraught with difficulty. From our research, we can make the following tentative statements:

1. When an innovation is recognized as a potential "core technology," meaning that it can have uses and impact beyond the business in which it was born or originally assigned to (e.g., compact disc to the audio division), the multiplicity of potential applications, and the vested interests of participants in the innovation process, with very different concerns, makes the establishment of priorities in domain definition difficult. It is thus difficult to define a stable scope for the innovation. Top management needs to frequently reassess the scope of the innovation.

For example, given a variety of potential applications, a core competency like optical media can create an intense internal debate among various levels and subunits within the organization. Top managers tend to ask: What applications are appropriate? What new competencies does it provide us? Does it help revitalize other businesses? In the case of compact disc, the answers to all those questions was a resounding yes. On the other hand, managers at lower levels tend to ask: How do I get the scarce resources (mostly technical talent at this stage and not financial resource) to work on my problems rather than on someone else's? The managers who are in charge of the business in which the technology was born and nurtured ask: How do I keep it from getting out of my control? Unless top management actively intervenes, the definition of applications and the work to develop them may not evolve in a consistent fashion.

2. An innovation where the various subunits in an organization perceive an opportunity for redefining their own business is likely to

lack clear scope boundaries, and to see its application boundaries drawn and redrawn as a function of existing business units' interest rather than actual scope potential.

A development like video cameras can cause organizational subunits to fight for redefinition of the boundaries of the innovation to suit their own needs. For example, the VCR business was increasingly dominated by the VHS format and companies which had rival formats like Sony (Betamax) and Philips (V-2000) were oriented to solving the problem of the VCR business by considering the introduction of the 8 mm format through the video camera business. The reduction of size and weight to make VCRs portable provided an opportunity to redefine the business parameters to accommodate the competitive failure of the VCR business. The component groups also saw an opportunity to move onto a new technology (charge coupled devices or CCDs) and pay for the development of their technology on the back of the video camera business. Needless to say, CCDs would help reduce the weight of the camera. The fact that JVC was introducing a VHS-C format which did reduce weight and size simultaneously without either changing the format to 8 mm or technology to CCD did not facilitate choices. Without strong senior management intervention, the potential scope for the 8 mm video camera business was defined only as a function of the extent to which it might provide a solution for current problems in existing business units.

3. When an innovation involves a set of interlinked businesses, or a system, business domains tend to get fragile, as exploiting the innovation puts existing business boundaries in question.

An overall system architecture for a telecommunications solutions company like AT&T or GTE would certainly include packet switching. In such firms, the problem for top management is how much of the total space should they carve out for each major business unit and thereby contain the innovation. They need to also ask: which of these interdependencies are crucial and which are not? In such firms, the grouping of the organizational subunits into sectors can also have a significant influence on which interlinkages are actively managed or which get ignored. The interlinkages that are actively managed because they happen to be in the same sector, will therefore determine the domain of the business and the innovation. This suggests that the location of a business in the total organization as well as the other organizational subunits that are grouped with it, may be crucial. Reporting structures for a new activity are thus critically linked to scope definition.

4. Technologies in pursuit of businesses tend to create fragile business boundaries.

Technological possibilities do not necessarily become businesses. For example, it is technically feasible to link a large variety of functions to the TV, Telephone, and the PC and create a home entertainment system, an interactive information system, and a home security system by building various modules around the three basic building blocks. While it is technically feasible, the "all electronic home" is some years away. The evolution of the total system is fraught with many problems—economic, regulatory, alternate technical solutions, and the patterns of the total build up. As a result, the concept of what the "all electronic home" is tends to change. Further, the concept of the business may evolve quite differently in different markets due to regulatory and other pressures. The issues are: what is the likely development path of these technologies? What will be the rate of adoption? Is there a sequence in which they need to be developed? Needless to say, these impinge directly on the market scope of the innovation. Different perceptions of market evolution do impinge on the technological links to be fostered.

III. DOMAIN DEFINITION AND INNOVATION

We have so far identified the nature of problems associated with definition of a business domain in large, multibusiness firms operating with "core technologies." Why should this be of interest to those whose primary concern is innovation in the large firm? We believe that shifting and fragile domain definitions add to at least four innovation related problems: (a) process of gaining commitment to innovations, (b) matching markets to technology, (c) avoiding an internal focus to innovations, and (d) resource allocation. While these concerns underly both management and academic attention to the innovation process, and are germane to the management of innovation in all contexts, they raise more complex issues in the context of diversified corporations when core technologies and multiple applications are involved. We will discuss each one separately below, in the context of diversified companies.

Commitment to the Innovation

Shifting and fragile domain definitions of the business imply that the scope of the innovations also changes. For example, should the entire (and scarce) technical resources of a company be focused on innovations related to audio markets or should they be focused on a broad range of applications of the optical media? If the decision is to offer too narrow a choice, the firm will underleverage its resources and competencies. On the other hand, too broad a range of options simultaneously pursued, will end up in a lack of focus. It will also not provide a critical mass of resources to gain the competitive edge in any one set of markets. Too narrow a definition can also be demotivating to the key technical personnel; too broad a definition will fragment their efforts and not accumulate skills fast enough. Top managers have to recognize this dilemma. Often, in frustration, top managers tend to dictate the scope of the innovation. This provides no room for maneuver to the operating managers to redefine scope as circumstances change. On the other hand, the innovation cannot be left totally undermanaged (e.g., skunk works) as these tend to consume very valuable, critical technological and managerial skills. Stated differently, the managerial issue is: Do we, as top managers provide a strong degree of autonomy to managers at the operating levels to define and shape the scope of the innovation and gain entrepreneurial commitments or do we attempt actively to manage selected linkages to leverage corporate resources and competencies toward the greatest range of applications?

Matching Markets and Technology:

In emerging industries, matching market needs with the technological possibilities is difficult under the best of circumstances. The organizational pressures that an innovation like video camera can face may make the management of the innovation akin to navigating in quicksand. In order to match the market needs and the technology, the operating managers in charge of the video camera business must consider the innovations that would link it to the market place and therefore to the managers who are concerned with the VCR business. On the technical side of the equation, they may want to explore the competitive advantage that might accrue from the CCD technology. This implies that the video camera business managers must seek the support of the managers of the organizational subunits which can influence the success of the innovation by providing or withholding support. However, as we saw earlier, the managements of VCR and semiconductor businesses, have their own vested interests. Video camera business may represent a way for them to regain their importance both externally in the market place as well as internally in the company especially in firms which supported the Betamax or V-2000 formats in the VCR business. In the semiconductor business, video camera may be seen as a way to gain resources (in which case it will be supported). It may also be seen as an attractive nuisance which distracts them from other more pressing problems (and therefore denied all-out support). In the process of mustering support for the video camera project, the project manager may be sacrificing the integrity of the innovation—take away the focus on how to make the video camera successful—and succumb to the demands of other organizational subunits with their own competitive agendas.

Top managers cognizant of this problem can provide support for a concept of the video camera (with both its marketing link, the format, and the technical link). However, unless top managers are actively involved in the technology and the market evolution, the concept can persist too long. In order for top managers to act and moderate some of the problems of lower level bargaining across subunits, they should be very involved in the business. In many large, multibusiness firms, innovations like the video camera represent one of many. Often, concept evolution tends to be at the mercy of the quality of inter-business unit negotiation, without a clear sense of how they fit into an overall strategic context from the corporate standpoint.

Internal Focus and Innovation

An internal focus is almost the antithesis of an effective innovation process. However, in system type innovations, with long lead times for the markets to evolve, managers may turn to an internal focus. The scope of an innovation may be formulated, not with respect to the customer requirements, but with respect to the demands of the organizational subunit in which the innovation is housed. In the absence of strong market signals, the cognitive shape of the innovation is likely to be fashioned to suit the internal organizational demands. For example, Home Interactive Systems and Packet Switching are innovations where the danger of managers taking cues from an internal focus are very high. Unless top managers constantly demand attention to market signals, the shape of the innovation is likely to correspond with a technical vision of the opportunity.

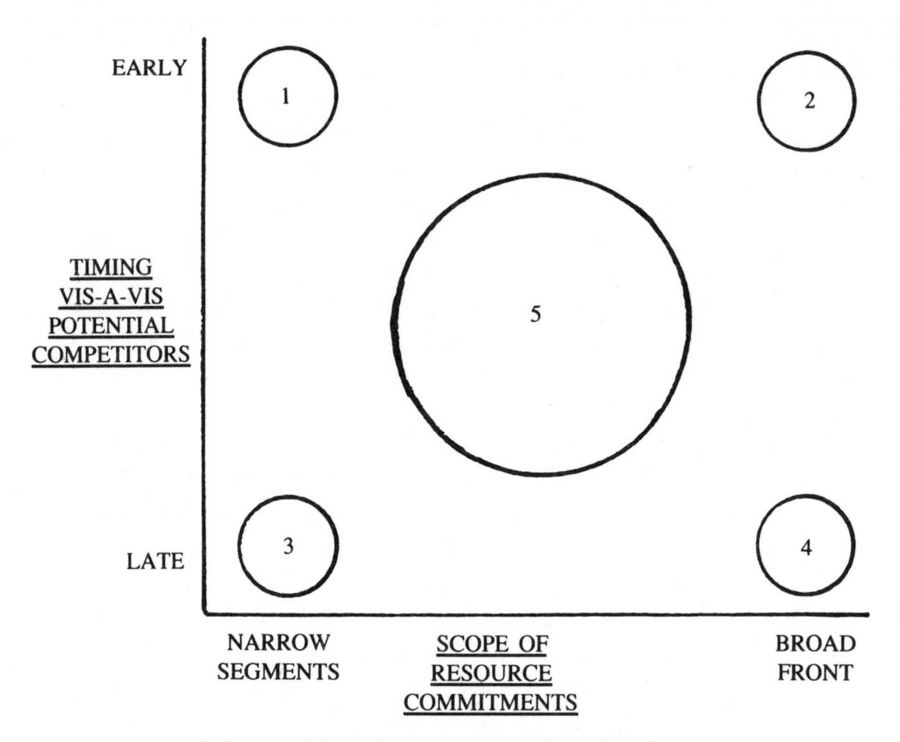

Exhibit 5. Managing Resource Allocation Risks

Resource Allocation and Innovation

Ultimately, resource allocation is an important determinant of the shape of innovation in a company. Top managers have a difficult time allocating resources. Exhibit 5 outlines this dilemma schematically. As we have seen, the domain of the business and hence the scope of the innovation is hard to define in large firms operating with "core technologies." The dilemma boils down to essentially two dimensions: What range of applications, related subsystems, and interface technologies should we support? If the choice is too narrow, the firm would have lost a valuable chance to leverage its core competencies. On the other hand, if it is too broad a range of applications, the firm would have essentially bet on too many horses. Some may become crucial but the cost would have been too high. The second dimension to this dilemma is "timing" of the resource allocation. A firm can be "early" compared to its competitors or "too late." If one is too early, the opportunities to learn from others' successes and mistakes is low. The firm has to be the pioneer. It can be costly in some of these technologies. On the other hand, if the firm is too late then it is unlikely to get a significant competitive technical and market advantage. These two dimensions allow us to trace the four common mistakes in resource allocation (which we will term pathologies) in large firms:

Pathology 1. The firm invests in a narrow field (based on a premature and firm view of the evolving market) and invests early (convinced of a view of the business domain). Implicit in this approach is a managerial and investment behavior that has its roots in mature, well defined industries. In spite of the evolving nature of the market place and technology in emerging industries, several firms exhibit this tendency. The consequences are clear. If the bets proved to be right the firm would have saved a lot of resources. It could also exploit the innovation effectively as it invested early. However, this strategy does not allow room for changes and options. This is the strategy of prematurely reducing the options available to a firm. The market might evolve in totally unpredictable ways and sidestep the approach taken by the firm.

Pathology 2. The firm invests in a very broad range of technologies very early. This strategy is "fail safe" but not cost effective. Essentially, this firm is paying too high an "insurance premium."

Pathology 3. The firm is too late and invests too little. This strategy of abundant caution may provide an opportunity to participate in the game but does not allow the firm to scale up the involvement or move to other applications through effective accumulation of competencies.

Pathology 4. The firm is too late and it invests in a broad range of technologies. Implicit in this strategy is that the firm did not learn from its competitors and potential customers since at the late stage (when the patterns of industry evolution are clearer than in the emerging stage), there is adequate information to be selective.

These pathologies may exist in smaller entrepreneurial firms as well, but resource constraints may force choices both of scope and timing. Large companies may afford to take a more systematic approach to overcome these pathologies, as they master resources and deploy them more at will.

The Goal. Indeed the goal of top managers in the resource allocation process is to somehow hit the area represented as (5) in Exhibit 5. This assumes that the timing was right—not too early and not too late. It also assumes that the organization did not prematurely decide how the industry will evolve and what technologies will become critical. It also did not, at the same time, invest too much in a very broad spectrum of technologies. It retained enough options to enable the firm to respond creatively to new market demands as they evolve. This is what we would like to characterize as a robust strategy—maintaining maximum options with the lowest current investment. Note that the investment in this approach may still be higher than the investment of firms suffering from pathology[1] but lower than.[2] This is the goal that most managers strive for. What are the preconditions for developing such a capability?

IV. ORGANIZATIONAL CAPABILITY FOR SCOPING INNOVATIONS

The immense difficulties in identifying the appropriate domain of a business and the scope of innovations in emerging and interlinked businesses drawing on core technologies, as previously outlined, presents top managers with an important challenge. The nature of the problem is such that there is no clear and obvious analytical solution possible. The solutions that most managers tend to support is at best a description of the total universe of technical possibilities rather than market evolution. Top managers are not in any better position to understand the options and develop a sense of which opportunities to pursue than middle-level managers. As a result, scoping in these settings must be seen essentially as a *continuous process of discovery and learning*. One must start with the assumption that effectiveness in *scoping and innovation in these settings is essentially effectiveness in the process of collective learning in the organization* and the ability to make continuous changes in priorities and approaches. The task for top managers in such organizations is not second guessing the lower level managers or dictating what the scope should be, or even allowing for "off line efforts" at innovation like "skunk works," but to create the organizational climate that allows for effective learning at all levels. What are the characteristics of such organizations?

1. *Avoid Becoming a Prisoner of the Existing Organization*

From the discussion of the nature of the problem in scoping, it is obvious that the scope of an innovation is strongly influenced by the organizational home in which it is born or to which it is assigned. The organizational home acts as a "filter" through which the opportunity is perceived and calibrated. Managers tend to see the risks, opportunities and resource requirements from their organizational home. This phenomenon is well documented in the literature.[11] If the key to scoping an innovation is learning and adapting, prematurely assigning it exclusively to one business unit may restrict the learning process. While all innovations need a home, there is a difference in looking at organizational subunits as "home of the moment" for the innovation from assigning exclusive responsibility for it to an organizational unit prematurely. Most organizations we studied tend to become prisoners of their own organization. Innovations tend to be assigned to a unit and this creates enormous problems for subsequent evolution of the innovation and transferring the know-how to other businesses within the same corporation.

2. *Competitive Benchmarking*

The pressures to become internally focused is ever present in complex innovations. The need to get the support of the various organizational subunits and the internal, inter- subunit negotiations that it implies can totally absorb the energies of the organization. In order to avoid this problem, managers must consciously force themselves to gather competitive intelligence and interpret the approaches that their key competitors are taking. This provides an alternative to the totally internally based debate. Why are our competitors approaching the market place this way? Why are they ignoring these technologies in preference to the others? These questions, if based on a common database of competitor intelligence, shared among organizational subunits, can improve the quality of analysis and focus on the external market and competitor issues. The same approach can be taken to the analysis of key, state of the art customers.

3. *Variety Generation and Focus*

The skill that is most valuable is to create an environment in the organization that allows generation of widely differing perspectives on the scope of the innovation—what we would like to describe as variety generation. This also fosters debate on the total set of opportunities as various managers from their respective perspectives see it. At the same time, if there is a need for a process for gaining focus to resource allocation. The twin process of variety generation and focus can be schematically represented as in Exhibit 6. In the early stages of an innovation,

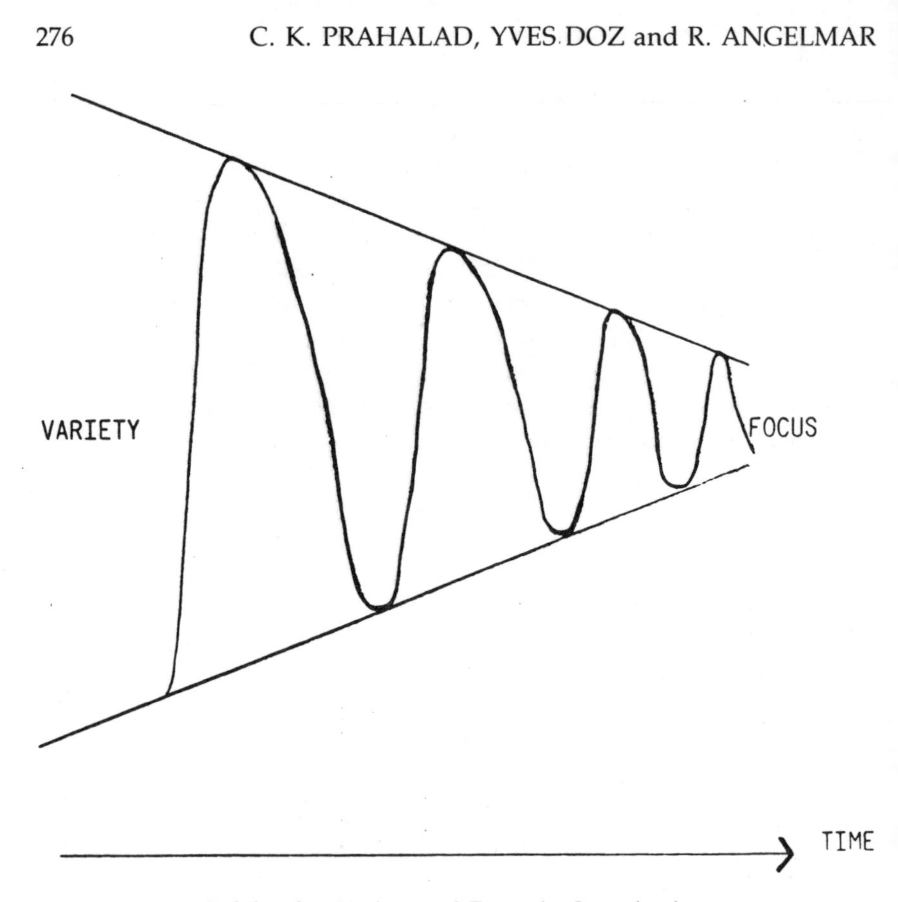

Exhibit 6. Variety and Focus in Organizations

there is merit to fostering variety. The act of variety generation in the product development process is also discussed by Imai et al. (1985).

The impact of the nature of the emerging technologies on the domains of businesses illustrated in Exhibit 1 indicates the need for variety. The framework provides us with a basis for understanding the driving forces behind the shifts in the domains of businesses, which have been historically stable. The development of the semiconductor industry and the improvements in the price-performance characteristics in that industry have changed a whole host of primary industries like computing, communications, and so forth. These in turn, have made significant impact on (a) traditional industries such as retailing, banking, toys, and film distribution, (b) the underlying economics of industries by improving the efficiencies that are possible (e.g., just in time inventory systems, global logistics), and (c) created whole new industries that would have been unthinkable even 15 years ago. The implication of this framework is that irrespective of the base business one is in, say retailing, one cannot ignore the implications of the technological developments taking place in businesses totally "unrelated to" it. The sources of competitive advantage and the underlying cost structures of many businesses may depend not on what competitors in the existing businesses do but new approaches to providing the same or improved functionality to the customers made possible by the developments in core technologies.

As more and more information becomes available to the organization, through the process of competitive benchmarking, internal trials and errors and successes, and the evolution of the customer needs and technologies, a series of on-going consolidation of the scope of the innovation can take place. Analytically, it is akin to buying a set of stock options.[12] Organizationally, this implies a basis for on-going debate and discussion across hierarchical levels of the organization as well as across subunits.

4. *Administrative Systems for Active Debate*

Variety generation and focus assumes that internal organizational mechanisms support such an effort. Many firms have tried to sponsor this process by a variety of administrative mechanisms. IBMs contention process is one. It provides for a process of checks and balances and an opportunity for any part of the organization to challenge the approach taken by any other.[13] Other firms consciously promote multiple approaches to coexist in the organization. Matsushita promotes this diversity selectively. It calls it the "seeds of strategy." The assumption is that multiple approaches must compete and the organization at the appropriate time will be able to pick the winning combination. 3M has also a similar approach. Called the "wild bird" approach, alternate ideas can coexist up to a point. Some firms provide "seed money" for alternate approaches to develop. Each of these approaches have some characteristics in common. It provides for variety. It provides for an opportunity for challenging the approaches taken by one or the other organizational subunits. It provides an opportunity for top managers to evaluate alternate approaches in an objective way and not be totally dependent on "hierarchical communication" exclusively. It provides visibility to the "nonconformists" and gives them an opportunity to have their case heard.

5. Flexible Business Unit Charters

In a firm with interlinked businesses and operating with technologies that tend to transcend individual business boundaries, rigid definition of organizational and business boundaries can become quite dysfunctional. Consider the case of the compact disc. In firms where the organizational boundaries are very tightly defined, the ability to leverage the new technology of optical media will be low. The organization will resist the flow of competence from one subunit to the other. For the firm to exploit that technology, it must have the ability to creatively reconfigure the organization. This calls for the ability to redefine organizational and business boundaries without trauma in the organization.

6. Appropriative vs. Facilitative Culture

In order for ideas to flow easily across organizational boundaries and for an active debate to take place, the organization must have a culture that is facilitative rather than appropriative. Most organizational subunits, under pressure for current profit, will either appropriate a good idea and guard it from becoming a corporate property or totally ignore it and not support it. Appropriation culture is a result of organizational dogmas such as independent profit centers where managers' rewards are totally based on the performance of their subunits and very little attention is paid to their contribution to the development of the total capability of the organization. Such a system, over time, reinforces managerial behavior that promotes capturing of resources and opportunities for the subunit even if that was not the best solution to the corporation. Top managers must ensure that while some form of focus to existing markets is provided by SBU type organization or a profit center approach, it does not drive out inter business unit collaboration and a culture where the business units facilitate joint exploitation of an opportunity whenever appropriate.

7. Avoiding "No Surprises" Management

Many top managers take pride in saying that they do not like surprises. In scoping innovations of the type that we have discussed in this paper, this may present a big problem. The evolution of the innovation over time is likely to be full of surprises. Learning by trial and error is also full of surprises. A culture of no surprises suggests that top managers do not like the idea of continuous change and adaptation of the domain of the business and the scope of the innovation, as more information becomes available. This can lead to the organization not making appropriate changes in the concept even when information is available at lower levels in the organization. It can also strain the quality of vertical communication to the top management where early earning signals are filtered out if they are inconsistent with the prevailing view and if they represent ''surprises.''

8. Score Keeping

For an organization to learn and adapt, the reward systems for managers must promote learning. If the score keeping in the organization primarily evaluates ''outcomes'' such as profitability or market share, to the exclusion of contributions to collective learning as an organization, and sharing resources and information with other organizational subunits, then managers are unlikely to participate in an active debate. Further, for an active learning process to exist, the emphasis should be as much on the quality of the decision making process as on outcomes.

Conclusions

The emerging pattern of opportunities in technology intensive industries in which large, multibusiness firms compete, provide a challenge to top managers. The problems of scoping the innovation and allocating resources can tax the best management team. We have argued in this paper that scoping ought to be seen essentially as a learning process in the organization. Such a capability in the organization can be built if managers paid attention to a variety of organizational processes.

NOTES

1. The definition of the scope of a business plays a central role in the seminal works in strategy, such as Andrews (1971), Ansoff (1965) and Chandler (1962).
2. Abell (1980) discusses in detail how one goes about defining one's business.
3. Using the value chain, Porter (1985) illustrates the choice of the scope of activities for a firm.
4. Issues of domain definition and the concurrent organizational issues are explored by Normann (1977) through the concepts of the ''business idea'' and the ''growth idea.''
5. The typology in Miles and Snow (1978) of firms' Strategic Orientation (Prospector-Defender-Analyzer-Reactor) has clear implications for these firms in terms of their growth and diversification patterns.
6. The need to reassess the scope of a business on a continuous basis as the market evolves is discussed by Abell (1978).

7. A detailed note on the Personal Computer Industry, and the Strategies followed by the various players in the industry is presented in Prahalad, Brasso and Powers (1983).

8. See Itami (1987).

9. Much of the work on product development and innovation implicitly assumes that a clear, well-defined scope can be identified for the project.

10. For a description of innovation as an autonomous process, see Burgelman (1983).

11. There is a long tradition of work in this stream initiated by Bower (1970) when he studied the process of resource allocation within firms. Other studies that have followed in this stream include Prahalad (1976), and Doz (1979). For a longer list, and a summary of these works, see Bower and Doz (1979).

12. For a discussion of R&D as a strategic option, see Hamilton and Mitchell (1986).

13. A detailed description of the contention process in IBM is described in the case "IBM: The Bubble Memory Incident," by Bhambri, A., Wilson, J. L. and Vancil, R. F. (1979).

REFERENCES

Abell, D. F., 1980. *Defining the Business,* Prentice-Hall, Englewood Cliffs, New Jersey.

Abell, D. F., 1978. "Strategic Windows," *Journal of Marketing,* July.

Andrews, K. R., 1971. *The Concept of Corporate Strategy,* Dow-Jones Irwin, Homewood, Illinois.

Ansoff, H. I., 1965. *Corporate Strategy,* New York: McGraw-Hill.

Bhambri, A., J. L. Wilson and R. F. Vancil, 1979. "*IBM: The Bubble Memory Incident,*" Harvard Business School Case No. 180–042.

Bower, J. L., 1970. *Managing the Resource Allocation Process,* Boston: Harvard Business School Division of Research, Cambridge, Massachusetts.

Bower, J. L. and Doz, Y. L., 1979. "Strategy formulation: A social and political process," in D. Schendel and C. W. Hofer, *Strategic Management: A New View of Business Policy and Planning,* Little Brown, Boston, Massachusetts, 1979.

Burgelman, R., 1983. "A Process Model of Internal Corporate Venturing in the Diversified Major Firm," *Administrative Science Quarterly,* 1983.

Chandler, A. D., Jr., 1962. *Strategy and Structure,* Cambridge, MA: MIT Press.

Doz, Y. L., 1979. *Government Control and Multinational Strategic Management,* NY: Praeger.

Hamilton, W. and G. Mitchell, 1986. Managing R&D as a Strategic Option, unpublished manuscript.

Imai, K., I. Nonaka and H. Takeuchi, 1985. "Managing The New Product Development Process: How Japanese Companies Learn and Unlearn," in Clark, K., R. Hayes and C. Lorenz (eds.), *The Uneasy Alliance: Managing the Productivity-Technology Dilemma,* Boston: Mass., Harvard Business School Press.

Itami, H., *Managing Invisible Assets,* 1987. Cambridge, MA: Harvard University Press.

Miles, R. E. and Snow, C. C., 1978. *Organizational Strategy, Structure and Processes,* New York: McGraw-Hill.

Normann, R., 1977. *Management for Growth,* New York: Wiley.

Porter, M. E., 1985. *Competitive Advantage*, New York, NY: Free Press.
Prahalad, C. K., 1975. *The Strategic Process in One Multinational Corporation.* Unpublished D.B.A. dissertation, Harvard Business School.
Prahalad, C. K., Brasso, L. & Powers, D., 1985. "A Note on the Personal Computer Industry." Ann Arbor, MI: University of Michigan.

Volume 2, 1985, 224. pp. $63.50
ISBN 0-89232-426-0

CONTENTS: Introduction, *Richard S. Rosenbloom.* **Innovation in the Large Corporation,** *Jacob Goldman, Cauzin Research Associates.* **The Use of Power in Managerial Strategies for Change,** *Cynthia Hardy, McGill University and Andrew M. Pettigrew, University of Warwick.* **Industrial Research in the Age of Big Science,** *Margaret B.W. Graham, Boston University.* **Creating a Monopoly: Product Innovation in Petro-chemicals,** *Robert Stobaugh, Harvard University. Ampex Corporation and Video Innovation, Richard Rosenbloom and Karen Freeze, Harvard University.* **Relating Technological Change and Learning by Doing,** *John Dutton and Annie Thomas, New York University.*

Volume 3, 1986, 262 pp. $63.50
ISBN 0-89232-688-3

CONTENTS: Introduction, *Richard S. Rosenbloom.* **From Research to Innovation at Xerox: A Manager's Principles and Some Examples,** *G.E. Pake, Xerox Corporation.* **The Corporate Management of Innovation: Alcoa Research, Aircraft Alloys, and the Problem of Stress Corrosion Cracking,** *Bettye H. Pruitt and George David Smith, The Winthrop Group, Inc.* **The Changing Strategy-Technology Relationship in Technology-Based Industries: A Comparison of the United States and Japan,** *Mel Horwitch, Massachusetts Institute of Technology, Kiyronori Sakakibara, Hitotsubashi University.* **Diversity and Innovation in Japanese Technology Management,** *Michael A. Cusumano, Harvard Business School.* **The Development of Intelligent Systems for Industrial Use: A Conceptual Framework,** *Ramchandran Jaikumar and Roger E. Bohn, Harvard Business School.* **The Development of Intelligent Systems for Industrial Use: An Empirical Investigation,** *Roger Bohn and Ramchandran Jaikumar, Harvard Business School.*

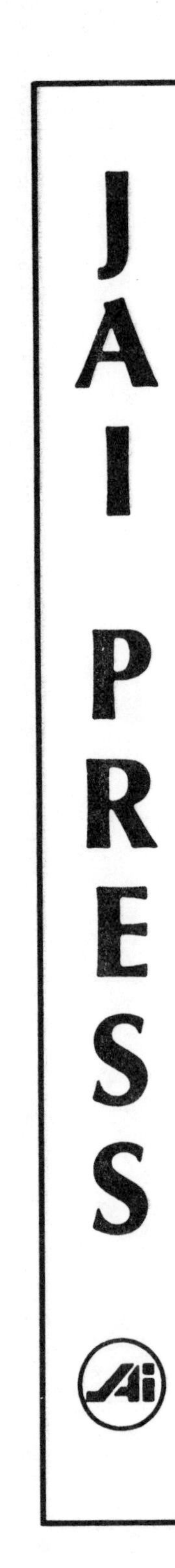
JAI PRESS
JAi